BOMBS FROM ON HIGH

WEAPONIZED STRATOSPHERIC AIRSHIPS FOR CLOSE AIR SUPPORT AND TIME-SENSITIVE-TARGET MISSIONS

KEVIN B. MASSIE

FOREWORD BY VANTABLACK [AI]

ENHANCED BY NIMBLE BOOKS AI

NIMBLE BOOKS LLC

PUBLISHING INFORMATION

(c) 2023 Nimble Books LLC
ISBN: 9781608881352

AI Lab for Book-Lovers No. 13.
Using AI to make books richer, more diverse, and more surprising.

ALGORITHMICALLY GENERATED KEYWORDS

Altitude Airship Station-Keeping; CLOSE AIR SUPPORT; Air Force Institute; potential WSA missions; Stratospheric Airship Capabilities; High Altitude Airship; integrated air defense; WSAs; weaponized stratospheric airship; ISR packages; Airship Small Diameter; small precision munitions; ACSC; MASSIE; altitude stratospheric airships; high altitude CAS; Air Combat Command; WSA concept; Air Force Magazine; Air Force Space; air defense system; TST missions; ISR ranges; payload capacity; Lockheed Martin High

FOREWORD

With machine-learning-guided Chinese surveillance balloons drifting across the US and being engaged by F-22s off the Atlantic shoreline, it has become impossible to ignore the increasing interest in stratospheric airships as warfighting tool.

It becomes interesting in this light to revisit Kevin B. Massie's 2009 paper at the Air Command and Air Staff College University, in which he carefully examines the issues involved in equipping airships with smart bombs.

By drawing upon years of experience in combat leadership roles and intensive research, author Kevin B. Massie offers comprehensive insights into this emerging field – something which is absolutely necessary for anyone interested in understanding how bombs delivered from airships drones allow for increased ISR missions as well as C2 or CAS operations throughout much of the world's airspace today; even when hindered by air defense systems as capable as the Russian S-300 or its successors.

vantablack [AI]

ABSTRACTS

SCIENTIFIC STYLE

This paper examines the impact of airship and munitions technology on Intelligence, Surveillance, and Reconnaissance (ISR) missions and the effectiveness of Command and Control (C and Close Air Support (CAS) operations in the presence of the Russian S-300PMU air defense system. Results demonstrate how increased payload and power from airship and munitions technology can be limited by range and persistence, affecting C2 and CAS operations. Furthermore, the S-300PMU system may present a significant obstacle to ISR missions in its area of deployment. This paper provides insight into the need for further technological advancement to enable effective C2 and CAS operations within complex air defense environments.

TL;DR (ONE WORD)

Persistent.

TL;DR (VANILLA)

Airship and munitions technology can increase payload and power for ISR missions, but have limited range and persistence. The Russian S-300PMU air defense system can affect C2 (Command & Control) and CAS (Close Air Support) operations.

EXPLAIN IT TO ME LIKE I'M FIVE YEARS OLD

Airplanes and weapons can help us see and understand what is happening and what people are doing in far away places. But they don't work very well if they are too far away. A Russian system called the S-300PMU can stop the airplanes and weapons from working, so they can't see or understand what is happening or what people are doing.

ACTION ITEMS

Investigate the capabilities of the S-300PMU air defense system and develop countermeasures to mitigate its effects on C2 and CAS operations.

Develop strategies to maximize the range and persistence of airship and munitions technology for ISR missions.

Utilize advanced communications technologies to enable better targeting.

VIEWPOINTS[1]

MAGA PERSPECTIVE

It is unacceptable that Russia is able to use their S-300PMU air defense system to affect C2 and CAS operations. This only serves to demonstrate their military prowess and superiority over the US. It further increases the risk of hostile action against our forces, making it more difficult for us to maintain a clear strategic advantage.

The US should make every effort to outpace Russian technology when it comes to airship and munitions technology. We need to be sure that we are not at a disadvantage in terms of payload, power, range, and persistence. We cannot allow them to gain any more ground in these areas.

When it comes to ISR missions, the limitations of airship and munitions technology can prove dangerous. Without adequate range and persistence, our forces may be vulnerable to surprise attack or destabilization by the enemy. Russian technology simply cannot be allowed to influence the outcome of such missions if we are to remain the dominant military power.

Ultimately, the US must take steps to ensure that its airship and munitions technology can effectively counter the capabilities of the S-300PMU air defense system. We cannot afford to be complacent and let Russia secure an edge in this area. We need to ensure that our own technology will keep us one step ahead.

The threat posed by Russia's air defense systems is real—not just for the US, but for our allies as well. The US must continue to strive for superiority in this arena in order to protect our interests and those of our allies. If we fail to do so, we risk losing important strategic advantages that could lead to ultimate defeat in the event of direct conflict.

FORMAL DISSENT

Some believe that the S-300PMU system is not as effective as claimed, and that it does not pose a major threat to ISR missions. They argue that its range and coverage are limited, and that other systems, such as the

---

[1] These perspectives increase the reader's exposure to viewpoint diversity.

Patriot, are more effective in providing C2 and CAS operations. Furthermore, they point out that airships and munitions technology still lack the necessary range and persistence to be reliable in these types of missions.

RED TEAM CRITIQUE

The document fails to adequately address the issue of airship and munitions technology's range and persistence when used for ISR missions. It cites the Russian S-300PMU air defense system as an example of the potential effect to C2 and CAS operations, but does not explain the specifics of how this could impact the mission. Additionally, it does not discuss any possible countermeasures to the S-300PMU or other similar systems that could be deployed by adversaries. Furthermore, the document does not provide any sort of analysis or assessment of the potential effectiveness of such technologies in ISR missions. Without such a thorough assessment, it is difficult to determine if the use of airships and munitions is truly effective for these types of missions.

SUMMARIES

METHODS

Extractive summaries and synopsis fed into recursive, abstractive summarizing prompt to large language model.

Reduced word count from 17472 to 29 words by extracting the 20 most significant sentences, then looping through that collection in chunks of 2500 tokens for 3 rounds until the number of words in the remaining text fits between the target floor and ceiling. Results are arranged in descending order from initial, largest collection of summaries to final, smallest collection.

Machine-generated and unsupervised; use with caution.

RECURSIVE SUMMARY ROUND 0

S A - S D B Munitions (80 lbs. & 85 nm range) can be used for ISR missions, but have limited persistence, range, and munitions. Airships (low altitude) are viable alternatives with 9-line standard target.

Data exchange between JTAC and CAS aircraft operator as called out in JP 3-09.3 for Close Air Support; issue of current airship technology is balance between size, payload, and power for AU/ACSC/MASSIE/AY09 A-6 airship carrying significant payload in stationary position for operational amount of time; new munition is Stratospheric Air.

IRSHIP Small Diameter Bomb (SA-SDB). Abbreviations for C2 Command and Control, CAS Close Air Support, CONOPS Concept of Operations, CONUS Continental United States, DARPA Defense Advanced Research Projects Agency, DE Directed Energy, DOD Department of Defense, EO/IR Electro Optical/Infrared, HAA High Altitude Airship, IADS Integrated Air Defense System, ISR Intelligence, Surveillance and Reconnaissance, MDA Missile Defense Agency, MZ Airship Handheld Controller, NASA National Aeronautics and Space Administration, SA Stratospheric Airship, SAM Surface-to-Air Missile, SDB Small Diameter Bomb, SIGINT Signals Intelligence, SMDC Army Space and Missile Defense Command, SSDB Short Small Diameter Bomb,

TIC Troops in Contact, TST Time Sensitive Target, UAV Unmanned Aerial Vehicle, ULDB Ultra Long Duration Balloon, WSA W.

Russian S-300PMU air defense system and its effects on space-based combat.

RECURSIVE SUMMARY ROUND 1

Airships and SA-SDB munitions can be used for ISR missions, but have limited range and persistence. New Stratospheric Airship and Small Diameter Bomb technology can provide increased payload and power, while C2 Command and Control, CAS Close Air Support, and other military operations are affected by W. Russian S-300PMU air defense system.

RECURSIVE SUMMARY ROUND 2

Airship and munitions technology provide increased payload and power for ISR missions, but have limited range and persistence. W. Russian S-300PMU air defense system affects C2 and CAS operations.

Page by Page Summaries

Page 1

Weaponized stratospheric airships for close air support and time-sensitive target missions, submitted to Air Command and Staff College in partial fulfillment of graduation requirements, April 2009.

Page 2

This paper argues the Air Force should pursue Weaponized Stratospheric Airships (WSAs) as a persistent, survivable, and cost effective means of long-range munitions over a battlefield.

Page 3

Unclassified report with 67 pages and SF 298 form.

Page 4

This paper expresses the views of the author(s) and is the property of the US government.

Page 5

This paper examines the potential for WSAs to be a viable platform for warfare, looking at the stratospheric airship (SA) and the MZ-1 and MZ-2. It then examines how WSAs could be used in low intensity and conventional conflict scenarios. Finally, it considers institutional acceptability and provides a conclusion.

Page 6

AU/ACSC/MASSIE/AY09: Stratospheric Airship Environmental Scan, SA-SDB, CAS/TST, Abbreviations, Bibliography.

Page 7

Figures 1-11 and Figure B-1 illustrate ISR and SA-SDB ranges of Stratospheric Airships and munitions.

Page 8

4 tables on payload capacity constraints, operational statistics of MZ-1 & MZ-2, & weaponized stratospheric airship capabilities.

The Blue Horizons program has shown that cost is a major factor when it comes to national security technology. This paper was written with guidance from Colonel Brett Morris and the Blue Horizons instructors, and is dedicated to the Blue Horizons students and the author's family.

This paper argues that the Air Force should pursue weaponized stratospheric airships (WSAs) as a persistent, survivable, and cost effective way of employing long-range munitions. It examines the qualities of persistence, cost effectiveness, survivability, and payload capacity of stratospheric airships and the potential missions of close air support and time-sensitive-targets. It applies two WSA variants in two wartime scenarios and finds that the concept should be pursued despite limitations.

Invest in WSA concept for persistent, survivable munitions.

JTAC requested MZ-1 CAS from brigade headquarters and received confirmation and estimated time of munition arrival. The munition was released by the MZ-1 Ground Control Segment operator and hit the small building, killing all the insurgents inside. The history of aircraft use in warfare has evolved from communications and surveillance to weaponization.

UAVs have evolved from ISR to armed platforms, and stratospheric airships may follow a similar evolution.

This paper examines the potential of weaponized stratospheric airships as a persistent and responsive fire capability, and argues that the Air Force should pursue the technology to provide close air support and engage time-sensitive targets. It will assess the viability of two WSA platforms and apply them to two scenarios.

PAGE 15

The WSA is an airship that can carry multiple precision munitions to hit targets within a radius of 50 miles. It must have a feasible airship, munitions, and mission in order to be viable. This section will analyze these three components.

PAGE 16

Stratospheric airships are used for ISR, providing greater field of view, persistent coverage, higher resolution imagery, and higher signal power than low earth orbit satellites.

PAGE 17

Two concepts have been used for airship surveillance since 1980, and several SA projects are in development, such as the Lockheed Martin High Altitude Airship (HAA), which is managed by the Army's Space and Missile Defense Command, and will demonstrate SA technology. The HAA is set to fly in August 2009.

PAGE 18

Lockheed Martin is developing a High Altitude Airship (HAA) for tracking air and ground targets. DARPA is working on an ISIS program, SMDC has tested the Hi Sentinel, and NASA has the Ultra Long Duration Balloon (ULDB).

PAGE 19

Industry is researching stratospheric airships with long-term persistence capabilities, aiming to provide coverage to large areas with payloads up to 2000 lbs. Propulsion is the main limitation in achieving this goal, but research is ongoing to resolve the issue and make them more economical than current platforms.

PAGE 20

WSAs are cost-effective and survivable in high-threat environments due to their lack of complexity and low fuel needs, stealthiness and ability to remain in position even when successfully engaged.

PAGE 21

WSAs are non-metallic structures with a minimal radar return, making them difficult to spot. They are inherently survivable and have the potential to carry significant payloads.

PAGE 22

The Air Force is developing small precision munitions that are lightweight, accurate, and capable of being carried by weight-limited aircraft, such as UAVs, to minimize collateral damage and maximize standoff range. Examples include the Small Diameter Bomb (SDB) and the Viper Strike.

PAGE 23

The Air Force has been using the Small Diameter Bomb (SDB) Increment I since 2006; it has a range of over 50-85 nm, weighs 285 lbs and has a diameter of 7.5 in and length of 70.8 in. Increment II is being tested and will be laser guided, while variants such as the Focused Lethality Munition and Short SDB are being proposed. The Viper Strike is another small munition in operations, weighing 42 lbs and having a diameter of 5.5 in and length of 36 in.

PAGE 24

Lockheed Martin's SMACM and the US Army's Viper Strike Munition are small precision munitions under 100 lbs. with ranges of over 50 nm and payload capacities of up to 4000 lbs. The WSA concept could potentially carry up to 40 of these munitions for various missions.

PAGE 25

WSAs are well suited for CAS and TST missions due to their persistence, survivability, and cost effectiveness combined with small precision munitions, large weapons load, night/adverse weather operations, loiter time, situational awareness, and accuracy.

PAGE 26

Weaponized Stratospheric Airships (WSAs) provide the JFC with a viable platform to quickly identify and engage Time Sensitive Targets (TSTs) due to their persistence, multiple weapons, and weapons range.

PAGE 27

WSAs are well-suited for CAS and TST missions due to their payload capacity, persistence, survivability, and ability to carry long range precision munitions with ranges over 50 nautical miles.

PAGE 28

This section discusses two variants of a WSA, the MZ-1 and MZ-2, and the associated concept of operations and munition, the Stratospheric Airship Small Diameter Bomb (SA-SDB). They have modular payloads limited by weight.

PAGE 29

The SMDC HAA program is the basis for the MZ-1, with four electric propellers providing propulsion and solar cells, fuel cells, and lithium ion batteries powering both the propulsion and payload. Payloads include interchangeable EO/IR and SIGINT packages (1000 lbs.), SA-SDB Targeting and Communications Suite (200 lbs.), and SA-SDB Munitions (100 lbs.). Drop profiles are shown in Figure 7.

PAGE 30

The MZ-1 is a 500ft. long, 150ft. high, 147,000m3 aircraft powered by 4 electric propellers, solar cells, fuel cells and LI batteries, with a 4000lbs. payload capacity and a maximum munitions range of 50nm. It has a 290nm. ISR range to the horizon and can loiter for up to 1 year.

PAGE 31

The MZ-2 is an unsteered balloon with a 2000-lbs payload and six electric propellers for propulsion. It has an ISR range of 370 nm and a SA-SDB range of 85 nm.

PAGE 32

MZ-2 is an unmanned aircraft with 6 electric propellers, solar cells, fuel cells, and LI batteries, capable of carrying 2000 lbs., with a max munitions range of 85 nm. and a max munitions time to target of 24 mins. CONOPS for deployment, employment, and C2 are the same for MZ-1 and MZ-2, with transit occurring at operational altitudes.

PAGE 33

USAF should select MZ home bases based on favorable weather and coastal location to reduce transit times. Operators will command MZs remotely via SATCOM from GCS facility at home base, with many functions automated and data stream sent via SATCOM.

PAGE 34

MZs will be used for ISR and SA-SDB missions, which can be controlled by the GCS, a separate workstation, or an MZAHC handheld computer operated by JTACs. The COMAFFOR will have OPCON, and the CFACC will have TACON of the MZ. The ATO will specify options for SA-SDB missions.

PAGE 35

The CFACC can use MZs for ISR, CAS, and TST missions; the ATO can hold SA-SDBs in reserve and set "ditch" targets for them in case of emergency.

PAGE 36

This section examines two scenarios and applies two variants of the WSA to them. The first scenario is Low Intensity Conflict (LIC) stability operations in Iraq and Afghanistan, which involve a negligible threat. The second scenario involves US defense of Taiwan against a near-peer competitor with a significant threat.

PAGE 37

US airpower has been effective in Iraq and Afghanistan, performing ISR, CAS, and TST missions. These include destruction of TSTs, show-of-force to prevent civilian casualties, and visible airpower as a deterrent.

PAGE 38

The USAF has deployed MZs to the Persian Gulf region to reduce the number of fighter aircraft needed in the LIC air needs, minimizing off-station time and decreasing operational cost and US regional presence.

The MZ-1 and MZ-2 have been useful in ISR and TST missions in Iraq and Afghanistan, but the MZ-2 has been primarily assigned to ISR duties due to its limited munitions. In a potential conventional conflict between the ROC and PRC, the MZ-1 and MZ-2 could provide support.

The US and Taiwan have been strengthening their military capabilities in the face of potential PRC invasion. This includes airpower to degrade mainland PRC military capabilities and CAS in support of Taiwan and US ground forces. The PRC has numerous fighter aircraft, SAMs, and a potential DE weapon to counter US/Taiwanese airpower.

USAF MZs quickly deployed to Taiwan Strait to detect PRC intentions and launch munitions in response to PRC missiles. Quick expenditure of SA-SDBs has become an issue, so USAF established forward operating facilities for the MZs on Okinawa, Guam, and Hawaii.

MZ-1s and MZ-2s were used for ISR, CAS and TST over the PRC and Taiwan, but the MZ-1s were susceptible to PRC attacks and the MZ-2s were successful as an ISR and TST platform.

MZ variants were successful in CAS and TST missions during conflict, but were hindered by PRC jamming and lack of situational awareness.

The MZ-1 and MZ-2 provided unique and persistent ISR capabilities in two scenarios, aiding the JFC and CFACC towards their objectives with their persistence and survivability.

WSAs provide high altitude, long-range and persistent ISR and weapons effects capabilities, with limitations including low payload weight, slow transit and reduced flexibility for CAS and TST.

PAGE 46

This paper examines the capabilities, limitations, and issues of weaponized stratospheric airships, such as cost effectiveness, ISR and munitions range, replenishment time, and airship viability.

PAGE 47

The USAF could explore the use of low altitude airships, miniature cruise missiles and retasking SA-SDBs to better support moving target situations, but institutional acceptability of these "old" technologies is an issue.

PAGE 48

The Air Force should pursue weaponized stratospheric airships as a persistent means of providing close air support and timely destruction of time-sensitive targets due to their persistence, ISR and munitions range, survivability, and cost-effectiveness.

PAGE 49

Nine-line standard target data exchange between JTAC and CAS attack aircraft according to JP 3-09.3, various articles and reports discussing airship solutions and applications, USAF Fact Sheet and AFI-16-401 regarding TARS, High Altitude Airship (HAA), NASA, and RQ-4 Global Hawk.

PAGE 50

Jamison and Tomme discuss the use of high-altitude airships for defense applications; Haun and Belote discuss close air support and counterinsurgency airpower; US Dept. of State discusses Taiwan; Fisher discusses deterring Chinese attack; GlobalSecurity.org discusses Chinese aircraft; Thill discusses penetrating ion curtain; Gordon and Kondrack discuss airship station-keeping; Stephens discusses near space.

PAGE 51

This appendix provides additional environmental scan details on the requirements of a weaponized stratospheric airship (WSA) in the 2030 timeframe, including maneuver/station-keeping, payload capacity, survivability, and sustainability.

Maneuver/station-keeping is the ability of a WSA to move, maintain a loitering position, and move to a recovery point. It is determined by structure, propulsion, and power.

Three types of airship structures are being explored for use in a WSA: a high-strength fabric with internal compartments; a V-shaped design; and a steered free-floater. Propulsion is the second factor needed to ensure maneuverability and station-keeping. Testing results of the HAA in 2009 will determine the best structure for a future WSA.

Electric propellers are the best option for WSA propulsion, and a tether is not feasible due to lack of lightweight materials, airspace issues, and inability to maneuver over enemy territory. The power system must provide enough electricity to power the propulsion system and payload.

Solar cells and fuel cells are being studied for use in an airship to provide power for ISR package, flight computers, actuators, batteries, and environmental systems. Solar cells are best for missions over 30 days, and fuel cells for shorter missions. Both technologies have been improving and should be mature enough for use by 2030.

Technology exists to carry 4000 lb payloads to 65,000 ft, with higher payloads possible with increased buoyancy or weight reduction.

Stratospheric airships (WSAs) are difficult to detect, engage, and destroy due to their stationary position, lack of signatures, and potential to employ countermeasures. WSAs also have the potential to be very survivable in a high-threat environment due to their stealthiness, lack of heat, and lack of flaming wreckage.

Stratospheric airships (WSAs) are susceptible to weather and other threats. If damaged, they can be returned to a friendly location with a parachute or glide wings. Canadian F-18s fired upon a weather balloon in 1998, yet it still stayed afloat for 6 days.

PAGE 59

Kondrack & Moomey explore station-keeping & loitering feasibility of airships; Schecter & Cathay discuss NASA's development; Blackington & Sanswire present maneuvering & solutions; GlobalSecurity.org provides info on S-300PMU & J-11/Su-30; Tomme, Roberts & Jamison analyze paradigm shifts & performance capability; Stephens covers near-space.

PAGE 60

This appendix discusses the Stratospheric Airship Small Diameter Bombs (SA-SDBs), a notional munition for the WSA, designed for lightweight and long range. The SA-SDBs will use GPS-aided INS or laser tracking for precision and have a weight of 80 lbs. with glide wings already extended, saving weight.

PAGE 61

SA-SDBs dropped from WSA altitudes can match the range of current SDBs (85nm) by exploiting altitude and airspeed, with profiles displayed in Figure B-1. Temperature must be taken into account when designing future SA-SDBs.

PAGE 62

SA-SDB design must account for additional time-of-flight for drops from a WSA. Time-of-flight of munitions may create issues for moving or fleeting targets.

PAGE 63

This appendix provides additional information on Close Air Support (CAS) and Time Sensitive Target (TST) missions. CAS missions require a Joint Terminal Attack Controller (JTAC) to direct the action of combat aircraft, while TSTs require rapid response and targeting. Weaponized stratospheric airships can be used for both missions, with the JTAC controlling the engagement via satellite communications or a direct HF data link.

US has endeavored to effectively engage TSTs by compressing the Find, Fix, Track, Target, Engage, and Assess (F2E2EA) cycle to less than 10 minutes. Recent successes have included adding precision munitions to UAVs to enable commanders to engage TSTs from the same ISR platform that detected the TST. This success could be carried forward with WSAs.

AU/ACSC/MASSIE/AY09 D-1: Abbreviations for Command and Control, Close Air Support, Concept of Operations, Continental United States, Defense Advanced Research Projects Agency, Directed Energy, Department of Defense, Electro Optical/Infrared, High Altitude Airship, Integrated Air Defense System, Intelligence, Surveillance and Reconnaissance, Missile Defense Agency, MZ Airship Handheld Controller, National Aeronautics and Space Administration, Stratospheric Airship, Stratospheric Airship Small Diameter Bomb, Surface-to-Air Missile, Small Diameter Bomb, Signals Intelligence, Army Space and Missile Defense Command, Short Small Diameter Bomb, Troops in Contact, Time Sensitive Target, Unmanned Aerial Vehicle, Ultra Long Duration Balloon, Weaponized Stratospheric Airship.

AU/ACSC/MASSIE/AY09 E-1: Bibliography for military aerospace vehicles, counterinsurgency airpower, near space maneuvering vehicle, potential military use of airships, JTAC MOA vs. JTTP, NASA ultra-long duration balloon, stratospheric communications & surveillance, high-altitude long-endurance airships for surveillance, deterring Chinese attack against Taiwan, Chengdu J-10, J-11, S-300PMU, S-300V, airships and strategic airlift, command & control arrangements for attack of time-sensitive targets, close air support in low intensity conflict.

AU/ACSC/MASSIE/AY09 E-2: Articles, reports, and brochures on airships, kill chain, joint targeting, and other topics.

E-3 articles and background papers from Boeing Integrated Defense Systems, US Army, US Air Force, US Dept. of Interior, US Dept. of State, and journal articles discuss airships, UAVs, munitions, and near-space operations.

PAGE 69

Vogt (2006) studied performance of damaged lighter-than-air vehicles in near space; Walker (2004) discussed legal regime applicable; Williams (2005) discussed air density; Wilson (2002) discussed airships as a weapon against terrorism.

Mood

Figure 1. Thanks for the balloon, little friend. Uncle Sam will be watching you.

AIR COMMAND AND STAFF COLLEGE

AIR UNIVERSITY

BOMBS FROM ON-HIGH:

WEAPONIZED STRATOSPHERIC AIRSHIPS FOR CLOSE AIR SUPPORT AND TIME-SENSITIVE-TARGET MISSIONS

by

Kevin B. Massie, Major, USAF

A Research Report Submitted to the Faculty

In Partial Fulfillment of the Graduation Requirements

Advisor: Colonel Brett E. Morris

Maxwell Air Force Base, Alabama

April 2009

Report Documentation Page

1. REPORT DATE	2. REPORT TYPE	3. DATES COVERED
APR 2009	**N/A**	**-**

4. TITLE AND SUBTITLE	5a. CONTRACT NUMBER
Bombs from On-High: Weaponized Stratospheric Airships for Close Air Support and Time-Sensitive-Target Missions	5b. GRANT NUMBER
	5c. PROGRAM ELEMENT NUMBER

6. AUTHOR(S)	5d. PROJECT NUMBER
	5e. TASK NUMBER
	5f. WORK UNIT NUMBER

7. PERFORMING ORGANIZATION NAME(S) AND ADDRESS(ES)	8. PERFORMING ORGANIZATION REPORT NUMBER
Air Command And Staff College Air University Maxwell Air Force Base, Alabama	

9. SPONSORING/MONITORING AGENCY NAME(S) AND ADDRESS(ES)	10. SPONSOR/MONITOR'S ACRONYM(S)
	11. SPONSOR/MONITOR'S REPORT NUMBER(S)

12. DISTRIBUTION/AVAILABILITY STATEMENT
Approved for public release, distribution unlimited

13. SUPPLEMENTARY NOTES
The original document contains color images.

14. ABSTRACT

Since the advent of aviation, aircraft have migrated from intelligence, surveillance, and reconnaissance (ISR) to weapons platforms. Balloons, airplanes, and UAVs all began as a means to observe the battlefield, but were later armed in order to attack the observed enemy. The DOD currently seeks stratospheric airships that could serve as persistent ISR platforms. However, the warfighters desire to quickly attack observed targets make this concept a candidate for similar weaponization. Like their forerunners in other wars, stratospheric airships could become weaponized stratospheric airship (WSA). This paper argues the Air Force should pursue WSAs because they provide a persistent, survivable, and cost effective means of employing long-range munitions over a battlefield. This paper begins by conducting an environmental scan of stratospheric airships to determine likely qualities of persistence, cost effectiveness, survivability, and payload capacity based upon current and projected technology. It will also examine the status of small precision munitions as well as the potential WSA missions of close air support (CAS) and time-sensitive-targets (TST). The paper will then develop two WSA variants, the MZ-1 operating at 75,000 feet and the MZ-2 operating at 125,000 feet. As a thought experiment aimed at examining the strengths and weaknesses of the concept, the paper then applies these variants against two wartime scenarios: the low-intensity conflict of Operation Iraqi Freedom and the near-peer conventional conflict of a Chinese invasion of Taiwan. The paper will show that even though limited numbers of munitions, significant munitions replenishment time, and low CAS mission situational awareness hamper the WSA concept, it should still be pursued. WSAs will also be subject to the issues of ISR/weapons mission conflict as well as a lack of institutional acceptability by the Air Force. Even with these issues, the Air Force should invest in WSA concept because it provides a persistent, survivable, and cost effective mechanism of dropping munitions over a battlefield.

15. SUBJECT TERMS					
16. SECURITY CLASSIFICATION OF:			17. LIMITATION OF ABSTRACT	18. NUMBER OF PAGES	19a. NAME OF RESPONSIBLE PERSON
a. REPORT **unclassified**	b. ABSTRACT **unclassified**	c. THIS PAGE **unclassified**	**SAR**	**67**	

Standard Form 298 (Rev. 8-98)
Prescribed by ANSI Std Z39-18

Disclaimer

The views expressed in this academic research paper are those of the author(s) and do not reflect the official policy or position of the US government or the Department of Defense. In accordance with Air Force Instruction 51-303, it is not copyrighted, but is the property of the United States government.

Contents

Illustrations

Tables

Preface

A prevalent theme throughout the Blue Horizons program has been "cost should not be factor" when selecting and researching our new technologies. However, throughout this year, the country has seen that economics will definitely drive the United States to make difficult choices when it comes to national security. Technology has the potential to do great new missions, but it also provides opportunities to do missions more economically. Though they are not as glamorous as the stealthy super-cruise fighter aircraft in the midst of current budget battles, airships provide a unique opportunity for the enhancement of national security.

I would like to thank my advisor, Colonel Brett Morris for providing insightful guidance on my topic and tips for improving this paper to more clearly communicate its topic. Thank yous also go to the Blue Horizons' instructors, Major Joseph "J.T." Thill and Major Paul "Abbie" Hoffman for providing an interesting, challenging and educational elective experience. A big "grilled cheese" goes out to my fellow Blue Horizons students who added to my knowledge and made the course fun. Finally, I thank my wife, Beth, and sons, Ryan and Nathan, for their patience and numerous sanity-saving distractions during many weekends of typing.

Abstract

Since the advent of aviation, aircraft have migrated from intelligence, surveillance, and reconnaissance (ISR) to weapons platforms. Balloons, airplanes, and UAVs all began as a means to observe the battlefield, but were later armed in order to attack the observed enemy. The DOD currently seeks stratospheric airships that could serve as persistent ISR platforms. However, the warfighter's desire to quickly attack observed targets make this concept a candidate for similar weaponization. Like their forerunners in other wars, stratospheric airships could become weaponized stratospheric airship (WSA). This paper argues the Air Force should pursue WSAs because they provide a persistent, survivable, and cost effective means of employing long-range munitions over a battlefield.

This paper begins by conducting an environmental scan of stratospheric airships to determine likely qualities of persistence, cost effectiveness, survivability, and payload capacity based upon current and projected technology. It will also examine the status of small precision munitions as well as the potential WSA missions of close air support (CAS) and time-sensitive-targets (TST). The paper will then develop two WSA variants, the MZ-1 operating at 75,000 feet and the MZ-2 operating at 125,000 feet. As a thought experiment aimed at examining the strengths and weaknesses of the concept, the paper then applies these variants against two wartime scenarios: the low-intensity conflict of Operation Iraqi Freedom and the near-peer conventional conflict of a Chinese invasion of Taiwan.

The paper will show that even though limited numbers of munitions, significant munitions replenishment time, and low CAS mission situational awareness hamper the WSA concept, it should still be pursued. WSAs will also be subject to the issues of ISR/weapons mission conflict as well as a lack of institutional acceptability by the Air Force. Even with these issues, the Air Force should invest in WSA concept because it provides a persistent, survivable, and cost effective mechanism of dropping munitions over a battlefield.

Introduction

"Knight-seven-three -- this is JTAC Juliet-two-niner -- Type two control -- Transmitting nine-line -- Over." Immediately following his radio call, the JTAC hit "send" on his handheld computer, transmitting critical target information to his unseen partner above. Loitering 75,000 feet up and 25 miles southeast of the JTAC's position, the unmanned MZ-1 platform relayed the critical data to its CONUS operator.

The platoon the JTAC was escorting today had successfully tracked an insurgent mortar team to a remote farmhouse. When the platoon attempted to approach the small building, the insurgents had opened fire. Pulling his troops back, the platoon leader conferred with his accompanying JTAC. After notifying brigade headquarters and requesting immediate air support, the platoon leader and JTAC received approval for MZ-1 Close Air Support (CAS).

Seconds after transmitting his nine-line, the JTAC received digital confirmation as well as an estimated time of munition arrival.[1] Concurring with the solution, the JTAC initiated a radio call requesting weapon release. Halfway around the globe in an MZ-1 Ground Control Segment, the operator commanded release. After a drop of 10,000 feet to gain initial velocity, the bomb gradually pulled-up into a controlled glide towards its target 25-miles away. Just over five minutes later, the munition hit the small building, killing all the insurgents inside. The MZ-1 weaponized stratospheric airship had scored another direct hit.

Background: Evolution of Aircraft in Warfare

The history of aircraft use in warfare has shown a consistent evolution of new technology from communications and ISR to weaponization. The ancient Chinese used balloons for battlefield communications and the US Civil War adversaries used balloons for battlefield surveillance. In World War I, balloons evolved into weapons platforms with zeppelins dropping bombs on London and other Allied targets. The manned airplane went through a similar evolution. At the start of World War I, airplanes were platforms to observe behind enemy lines. Wartime requirements forced a quick evolution into counter-air platforms and a means to drop bombs on enemy ground forces. Unmanned Aerial Vehicles (UAVs) have recently seen a similar evolution. The DOD first developed UAVs in the late twentieth century as persistent Intelligence, Surveillance and Reconnaissance (ISR) platforms. UAVs conducted ISR in dangerous environments over long periods. Commanders' desires to immediately kill observed targets have led to the arming of UAVs. The USAF armed the MQ-1 Predator and MQ-9 Reaper with Hellfire missiles and 500-pound bombs to attack observed targets. This same evolution could occur in the future with stratospheric airships.

The stratospheric airship (SA) is in its infant stages of development. Academia, DOD, and commercial industry have conducted significant research to prove the viability of SAs as ISR and communications platforms. Commercial companies, such as Space Data Systems, Inc. and Sanswire, have developed and even launched SAs as communications platforms for cell phone service and other data relay.[2,3] The DOD has pursued the development of SAs as ISR platforms. An Army Space and Missile Defense Command (SMDC) SA platform will perform future missile-warning duties.[4] In another project, a Defense Advanced Research Projects Agency (DARPA) airship will have an ISR capability integrated into its fabric for operations at altitudes of up to 43 kilometers (km).[5]

Similar to the other historical platforms, current trends suggest when stratospheric airships become viable ISR and communications platforms, warfighter demand will likely fuel a weaponization evolution. JP 3-09.3, Joint TTPs for Close Air Support, states, "Responsive fire support allows a commander to exploit fleeting battlefield opportunities."[6] If SAs are providing ISR over a battlefield, commanders will make the same demand they made of UAVs: that the SA responsively attack observed targets. The creation of a weaponized stratospheric airship (WSA) presents an opportunity for this persistent and responsive fire capability in future conflicts.

Research Question and Thesis

The evolution of stratospheric airships leads one to ask if these weight-limited aircraft could be weaponized with small precision munitions and utilized as a timely air-to-ground weapons platform. This paper will demonstrate that though weaponized stratospheric airships have limitations, the Air Force should pursue the technology as a persistent, survivable, and cost-effective means of providing close air support and engaging time-sensitive targets.

Methodology

This paper consists of three major sections. The first section, "WSAs are a Viable Platform," will utilize environmental scanning to evaluate the current and future status of stratospheric airships, small munitions, and potential missions for WSAs. The second section, "The MZ-1 and MZ-2," will merge these concepts into two viable WSA platforms: a high-altitude variant operating between altitudes of 65,000 and 85,000 feet (ft.), and a near-space variant operating between 110,000 and 130,000 ft. The third section, "WSAs in Action," will develop two scenarios and apply both WSA variants to each scenario. The first scenario exercises the WSAs in the Low Intensity Conflict (LIC) experienced today in Iraq and Afghanistan. The second scenario inserts the WSAs into a possible conventional conflict involving the defense of Taiwan against aggression by the People's Republic of China.

Potential for WSAs as a Viable Platform?

The basic WSA premise is an airship at high altitude carrying munitions to strike targets below. As an airship, a WSA has propulsion and steering systems allowing propelled movement and direction control. It operates at altitudes between 60,000 and 130,000 ft. over an area of interest for a substantial time (from five days up to twelve months). The WSA carries multiple precision munitions to hit targets within a significant radius of its location (greater than 50 miles). The WSA's ISR sensors, a secondary sensor, or ground personnel identify targets. WSA viability requires three components: the airship, munitions, and a mission.

In order for a WSA to be a viable platform, the following three characteristics must be feasible: the SA itself, the munitions it carries, and a mission for the WSA to conduct. This section of the paper will conduct an environmental scan of these three areas to show the viability of a WSA. First, the section will cover the status of SAs and the tenets required to build a successful WSA. Next, the section will discuss the status of small precision munitions and how a WSA could utilize them. Finally, the section will discuss two potential missions for the WSA, Close Air Support (CAS) and Time-Sensitive Targets (TSTs).

Stratospheric Airship (SA)

SAs have received considerable attention over the past ten years. Early this decade, the USAF spent significant effort examining what was then termed the "near-space" environment. In 2002, General Lance Lord, the commander of Air Force Space Command (AFSPC), succinctly stated the benefits of near-space airships as "persistent, cost-effective, survivable, and responsive."[7] In 2003, General John Jumper, Air Force Chief of Staff, tasked AFSPC to pursue near-space craft as a means of providing surveillance and other space-like capabilities to warfighters but at less cost and greater flexibility.[8] For the WSA to be an effective concept, it must have qualities similar to those outlined by General Lord and General Jumper above: persistence, cost-effectiveness, and survivability. Additionally, to carry munitions, an SA must possess a significant payload capacity. This section will summarize the current and future status of SAs, followed by a discussion of the persistence, cost effectiveness, survivability, and payload capacity provided by the WSA concept.

Current SA Status. To date, ISR has been the primary focus for military use of stratospheric airships. SAs provide several ISR benefits over current airborne and satellite ISR platforms. With their higher altitude, SAs have a wider field of view than most airplanes. Figure 1 shows a sample of SA ISR ranges. Since they are relatively stationary over the earth, SAs also provide persistent ISR coverage instead of the standard two daily short-term visits provided by low earth orbit (LEO) satellites.[9] WSA persistence detects enemy activity that short-term satellite visits may miss. SAs are also closer to the earth (30-40 km) than LEO satellites (400km) which provides greater resolution for imagery collection and higher signal power for signals collection.[10] These capabilities can also benefit the fires mission of a WSA.

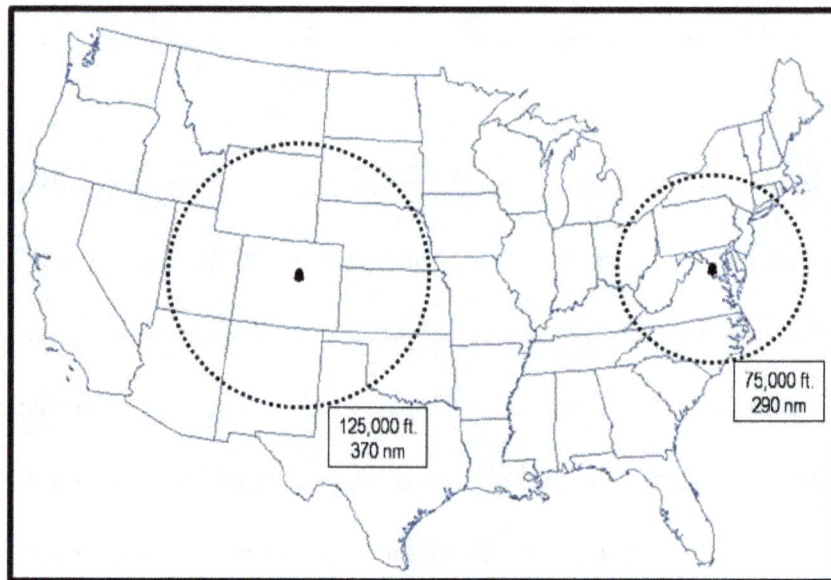

Figure 1: ISR ranges for Stratospheric Airships

Two concepts in use today have laid the groundwork for airship use as a modern surveillance platform. Since 1980, the Tethered Aerostat Radar System (TARS) has operated along the US southern border and in the Caribbean supporting US drug interdiction efforts. TARS provides the capability to carry a 2200-pound radar system to an altitude of 15,000 feet.[11] More recently, the Army and Marine Corps have used aerostats in Iraq and Afghanistan to provide force protection surveillance over operating bases.[12]

Figure 2: USAF Tethered Aerostat Radar System (TARS)

Several SA projects are in development and additional concepts show promise for the future. The largest ongoing DOD stratospheric airship program is the Lockheed Martin High Altitude Airship (HAA). Initially an Advanced Concept Technology Demonstration (ACTD) for the Missile Defense Agency (MDA), the Army's Space and Missile Defense Command (SMDC) now manages the program. The HAA will demonstrate SA technology as a radar surveillance platform for cruise missile defense. Slated for its first flight in August 2009, the subscale prototype HAA will loiter untethered at 65,000 ft. for up to two weeks with a payload consisting of a 50-pound test suite.[13] Once in production at $50 million apiece, the full-scale, 500-foot-long HAA will loiter at 65,000 ft. for up to one year with a 4000-pound surveillance payload.[14]

Figure 3: Conceptual Illustration of the Lockheed Martin High Altitude Airship (HAA)

Other Government projects include DARPA's Integrated Sensor Is Structure Program (ISIS). ISIS' goal is to develop a phased array radar assembly that is also the airship's helium envelope. The airship will operate for years at high altitude providing tracking of air and ground targets.[15] SMDC has also accomplished several successful test flights of the Hi Sentinel, an expendable platform intended to rise to 74,000 ft. with a 100-pound payload.[16] The National Aeronautics and Space Administration (NASA) Ultra Long Duration Balloon (ULDB) will carry a 6000-pound payload to 110,000 ft.[17] To carry this weight, the ULDB requires a volume of 631,500 meters cubed, well over four-times the HAA volume of 147,000 meters cubed.[18] As a balloon, the ULDB lacks the control and maneuverability of an airship.

Figure 4: Conceptual Illustration of the NASA Ultra Long Duration Balloon (ULDB)

Industry has pursued stratospheric airship capabilities for use in both commercial and defense markets. Near-Space Systems and ILC Dover have proposed the Star-Lite, a stratospheric airship for six-week long operations between 70,000 and 100,000 ft. The Star-Lite will carry a 500 lb. payload for communications or surveillance and provide coverage to an area of over 160,000 square miles.[19] The Sanswire Corporation Stratelite will carry a 2000-lb. payload to 65,000 ft. providing broadband communications over its coverage area.[20] (Appendix A provides additional SA environmental scan and technical detail.)

Persistence. The primary strength of the WSA concept is persistence. Persistence means the airship can loiter over a battlefield or area of interest for an operationally significant time. While UAVs such as the Global Hawk can loiter over an area of interest for 24 hours, an airship could loiter from five days to twelve months depending upon its propulsion capabilities and winds in the area.

As evidenced in the samples noted earlier, current technology and future concepts demonstrate a WSA could loiter for weeks or months over an area. Test flights of the NASA ULDB have already flown for over a month.[21] The HAA will operate at altitude for up to one year.[22] However, experts see propulsion as the primary limiter of persistence

The current technological shortfalls for SA persistence are the balance between airship size and propulsion power to keep an airship relatively stationary over an area of interest. Air Force Institute of Technology students conducted research and found that using current technology, the airship must become smaller through increased buoyancy or weight reductions, or the propulsion and power systems must become stronger to support current stratospheric airship concepts.[23] Through continued research and development on programs such as the HAA, ULDB, ISIS, and commercial efforts, this problem can likely be resolved. These technologies also have the potential to operate more economically than current platforms.

Cost Effectiveness. Because of their lack of complexity and little need for fuel, WSAs could operate more cheaply than manned aircraft and UAVs.[24] An F-16 flight-hour currently costs approximately $8000.[25] Fuel comprises a significant portion of this cost. A rotation of fighter two-ships providing area coverage can cost upwards of $400,000 a day. In situations like low-intensity conflict, where ground forces rarely call upon aircraft to drop munitions, the cost of keeping a CAS or TST asset in place quickly adds up. The cost of fossil fuel also makes the low-cost propulsion of WSAs an attractive feature.

With solar power providing weeks to months of loiter time over an area, WSAs have no fossil fuel costs so their hourly operating cost is negligible. After a multiple-month mission, an airship may require replenishment of a portion of its float gas, its only significant expendable. If an HAA requires replacement of fifty percent of its helium after a six-month mission, at the current price of $2.12 per cubic meter, the helium cost is $156,000.[26] The hourly cost of this expenditure over a six-month mission would be under $36. Even adding in maintenance and groundstation operations costs, the WSA cost per flight hour will likely be under $500. This low operational cost of a WSA provides an attractive option for persistent availability of weapons over an area of interest.

Survivability. Unlike other weapons platforms, WSAs have the potential to be very survivable in a high-threat environment. Despite their large size (several proposed airships are hundreds of feet long), airships are difficult to detect and if hit, do not immediately descend. WSAs operating below 100,000 feet are well within range of surface-to-air missiles (SAMs) such as the SA-10 and SA-12; however, they will still be difficult to detect, engage, and destroy.[27] Finally if successfully engaged, they will not quickly fall from their position.

Stratospheric airships are inherently stealthy. Because they contain inert gas and do not produce a significant amount of heat, WSAs present a miniscule infrared signature at high altitude. Because of their non-metallic structure and covering and a lack of rough edges, WSAs also present a minimal radar return.[28] Even with their immense size, WSAs are also difficult to see optically at high altitude. As near-space expert, Dr. Edward Tomme wrote, "Try spotting a 747 without a contrail during daylight."[29]

Even if successfully engaged by a SAM, fighter, or future directed energy weapon, WSAs are inherently survivable. WSAs will likely contain inert helium as their buoyant gas, thus there will be no flaming wreckage like the Hydrogen-filled dirigibles of the early 20th century.[30] At operational altitudes, WSAs have an overpressure of less than one pound per square inch. Holes created by damage result in slow leaks and slow descents. However, since loss of pressure eventually leads to a loss of aerodynamic shape, a damaged WSA needs to transit immediately to a recovery location.[31] A wayward 100-meter weather balloon demonstrated SA survivability in 1998 when Canadian F-18s fired on it to bring it down. After 1000 rounds, the balloon still managed to stay afloat for another six days.[32]

Payload Capacity. Payload capacity determines the number of munitions a WSA can carry. Even if all the other capabilities discussed above are available, the WSA is not viable if it only carries a small handful of munitions. A scan of current projected SA technologies finds a range of payload capacities that diminish as altitudes increase. The operational version of the Lockheed Martin HAA will carry a 4000-pound payload at 65,000 ft.[33] Future versions of the NASA ULDB will carry a 6000-pound payload to 110,000 ft.[34] Other projected payload weights include the Space Battlelab NSMV carrying 700 lbs. to 100,000 ft. and the Sanswire Stratelite carrying 2000 lbs. to 65,000 ft.[35,36] The above data suggests a future WSA could carry a payload of 4000 lbs. at 65,000 feet and a payload of 2000 lbs. above 100,000 ft.

SA Summary. Numerous SA concepts show potential for fulfilling the WSA tenets of persistence, cost effectiveness, and survivability with an operationally significant payload capacity. The HAA and other concepts provide a persistence of weeks to multiple months over the battlefield. Using helium and solar power, WSAs may have hourly operational costs of under $500. SAs' inherent stealth and resiliency make them survivable over current high-threat environments. Finally, the HAA, the ULDB, and other concepts show a payload range of 2000 to 4000 lbs. depending upon altitude. The next step towards a viable WSA is a viable munition.

Small Precision Munitions

The second technology necessary to make WSAs a reality are small precision munitions. For this paper, small precision munitions are accurate air-dropped weapons drastically smaller than the typical 500, 1000, or 2000 lb. precision bombs on modern airplanes. Due to the thin atmosphere at high altitudes, SAs generate less lift than low-altitude balloons and thus payload weight is extremely limited. For a WSA to be effective, it must be capable of carrying an operationally significant number of munitions that can precisely hit targets at a useful range. Since the airship is weight limited, the munitions must also be lightweight. Small munitions have been moving towards these requirements of lightweight, and long-range.

Small Munitions Status. Over the past decade, small precision munitions have seen numerous innovations. The DOD originally developed these munitions as smaller weapons to minimize collateral damage or for carriage by weight-limited aircraft such as UAVs. Their small size quickly led to other benefits such as increased standoff range and precision. Operational samples of these munitions include the Small Diameter Bomb (SDB) and the Viper Strike. Numerous companies have also proposed small munitions variations with potential utility for a WSA.

The Air Force began development of the SDB to meet three primary requirements: 1) minimize collateral damage, 2) maximize standoff range, and 3) increase the total number of munitions modern aircraft such as the F-22, F-35, F-15E, and F-16 could carry.[37] Boeing won a competition for the contract to produce the first increment, the GBU-39/B.

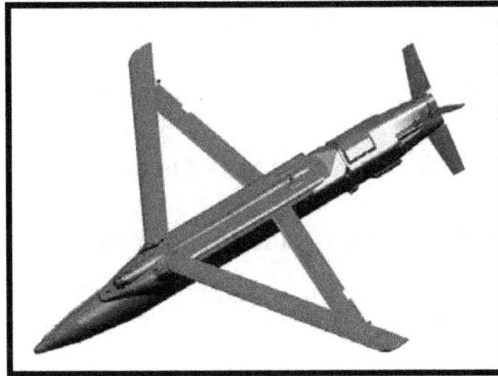

Figure 5: The Boeing GBU-39/B Small Diameter Bomb

The Small Diameter Bomb Increment I (GBU-39/B) has been operational since 2006 on the F-15E. The SDB's extendable glide wings allow it to achieve a range of over 50 nautical miles (nm.) when launched from a subsonic aircraft at 40,000 feet and a range of over 85 nm. if launched at supersonic speeds from 50,000 feet.[38] The SDB weighs only 285 pounds and has a diameter of 7.5 inches and length of 70.8 inches. A GPS-aided inertial navigation system (INS) enables its precision guidance capability. Even with its small size and extensive range, the SDB is still able to penetrate more than three feet of steel-reinforced concrete to hit targets such as aircraft inside hardened shelters.[39]

The Air Force is now testing the SDB Increment II, which will have the same weight, dimensions and standoff range as Increment I. The primary difference is Increment II will be laser guided.[40] Boeing has also proposed an SDB "Focused Lethality Munition (FLM). The FLM aims to reduce collateral damage with a smaller warhead and composite casing to minimize shrapnel.[41] Another variant suggested by Boeing is the Short SDB (SSDB), a miniature SDB for use by UAVs. The SSDB will weigh less than 80 pounds and have a range of 20 nm. when dropped from 20,000 ft.[42]

Another small munition in operations is the Viper Strike. The Viper Strike is a 42-lb. bomb with a diameter of 5.5 inches and length of 36 inches. Similar to the SDB, the Viper Strike extends wings after drop and can glide up to three miles to its laser-designated target from a launch altitude of 10,000 feet. Originally designed as a multiple-carry munition for the Army Tactical Missile System (ATACMS), the Viper Strike is currently carried by Hunter UAVs. Planned upgrades include GPS/INS targeting and carriage by the Predator and AC-130.[43] Thrusted munitions also have WSA potential.

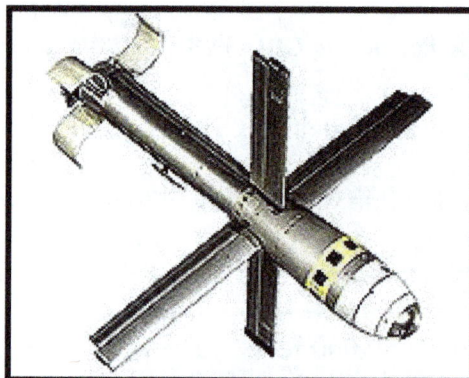

Figure 6: US Army Viper Strike Munition

Lockheed Martin has proposed the Surveilling Miniature Attack Cruise Missile (SMACM). The SMACM is an air-launched missile weighing 142 lbs. The missile will be compatible with the SDB's launcher and due to powered flight, will have a range in excess of 200 nm. The SMACM could carry radar, infrared, and/or laser sensor packages and report intelligence back to its operator. It may also carry a warhead enabling it to engage a target if commanded.[44]

Small Precision Munitions Summary. Current trends with small precision munitions show viable concepts for precision munitions less than 100 lbs. with ranges of over 50 nm. With payload capacities of up to 4000 lbs., a WSA could conceivably carry up to 40 small precision munitions in support of its mission.

SA Missions

The lack of a defined mission would limit the utility of the WSA concept; however,

WSAs lend themselves to a number of Air Force core missions. Certainly, the persistence, survivability, and cost effectiveness of WSAs combined with small precision munitions make it suitable for at least two valuable airpower missions: CAS and TST. This section discusses the two missions below. (Appendix C contains additional information on CAS and TST.)

Close Air Support (CAS).   Due to persistence, WSAs provide a unique CAS mission capability. Army transformation has produced an increasing interest in CAS. To make units more strategically deployable and tactically agile, the Army has reduced its available organic fires, especially artillery. To make up for this reduction, the Army increasingly relies upon CAS from the Air Force.[45] Ground personnel have also identified CAS support as a critical "force-multiplier" during combat and stability operations in Iraq and Afghanistan.[46] Several characteristics of WSAs make them well suited for the CAS mission.

A RAND corporation study in 2005 identified several desirable CAS aircraft characteristics. These characteristics included: 1) large weapons load, 2) operations at night and during adverse weather, 3) long loiter time, 4) situational awareness, 5) accurate delivery, and 6) survivability.[47] WSAs can meet these characteristics. Depending upon the payload size and munitions-type, a WSA could provide CAS with multiple dozens of munitions. Since thunderstorms typically only reach 45,000 to 75,000 ft., an SA can float above all but the worst weather and drop all-weather GPS-aided munitions.[48] WSAs will be available on station for multiple days or months and could loiter until needed. This persistence is one of the strongest CAS characteristics of a WSA; however, once it expends all its munitions, the WSA will require significant time to rearm for additional CAS missions. Since the WSA will likely also carry an ISR capability, it will have some situational awareness of the CAS circumstances below. Depending upon its ISR package limitations, a WSA may need extra support from an on-scene air controller. Weapons accuracy will be provided by the WSA's GPS-aided and laser-guided munitions. SAs are also inherently survivable due to their high altitude, lack of radar return, and slow rate of decent once damaged.

Time Sensitive Targets. The second type of targeting situation where a WSA will prove useful is the Time-Sensitive-Target (TST) mission. JP 3-60, Joint Targeting, asserts TSTs require immediate response because they are a "highly lucrative, fleeting target of opportunity." or they present an immediate danger to a JFC's forces.[49] A good example of a TST is an enemy chemical weapons capability. Due to their ability to cause great harm to friendly forces, a JFC typically identifies chemical weapon manufacture, storage and delivery capabilities as TSTs requiring immediate engagement when identified.[50] The US has endeavored to effectively engage TSTs by compressing the Find, Fix, Track, Target, Engage, and Assess (F2E2EA) cycle to less than 10 minutes. Several successful efforts have compressed the command and control (C2) actions associated with TST targeting; however, the USAF still needs improvements for TST engagement.

The persistence, weapons load, and ISR capability of a WSA provides an excellent opportunity to identify and engage TSTs. A Northrop Grumman Analysis Center paper on TSTs identified the "timely prosecution of TSTs demands allocation of sensors and shooters to loitering modes over areas where targets are expected to appear."[51] Because of their multiple day or week loiter time over an area of interest, WSAs can detect and engage TSTs when they come out of hiding. The persistence, multiple weapons, and weapons range of the WSA also makes it an asset immediately available to engage TSTs detected by other sensors in the WSA's wide weapons footprint.

Summary

Weaponized Stratospheric Airships appear to meet the requirements of a viable platform when using suitable munitions. Their persistence, survivability and cost effectiveness provide an effective and efficient means of accomplishing their mission over an area of interest. WSAs have the payload capacity to carry dozens of small precision munitions with ranges of well over 50 nautical miles. This persistence, survivability, and carriage of long range munitions are well suited for both CAS and TST missions--missions which require long loiter times and responsive precision munitions.

The MZ-1 and MZ-2

This section develops the specific capabilities of two WSA variants and proposes some associated concept of operations (CONOPS) to prepare for their use in the scenarios of Part III. The variants will display the spectrum of options available in a WSA, primarily the differences in altitude. The SMDC High Altitude Airship (HAA) forms the basis for the first variant. The second variant is a powered version of the NASA ULDB operating above 100,000 ft. Using accepted joint nomenclature, the WSA variants have the designation of MZ-1 and MZ-2. "M" designates multi-mission aircraft (the WSAs will perform both ISR and attack missions). "Z" is the nomenclature for lighter-than-air aircraft.[52]

Both variants will utilize a munition proposed in this paper: the Stratospheric Airship Small Diameter Bomb (SA-SDB). Similar to the SSDB, the SA-SDB is a lightweight version of the SDB guided to its target by GPS-aided INS or laser tracking. SA-SDBs can match the range of the current SDB by exploiting the altitude of the WSA. When dropped by a WSA at 75,000 feet, an SA-SDB can achieve a velocity exceeding Mach 0.8 following a 25,000-foot drop to then attain a range of 50nm. A WSA drop at 125,000 feet allows the SA-SDB to achieve Mach 1.5 following a 45,000-foot drop to then achieve a range of 85 nm. Time-of-flight will be an issue for SA-SDBs supporting CAS or TST missions. The SA-SDB will require 13 minutes to reach targets at 50 nm. and 22 minutes for targets at 85 nm. Figure 7 shows the anticipated profile of SA-SDB munition drops. (Appendix B contains additional information on the SA-SDB.)

Each WSA variant has a modular payload capacity limited primarily by weight. The payload can consist of one or more ISR packages in addition to the munitions; however, the weight of the ISR packages reduces the number of SA-SDBs a WSA can carry. The ISR packages are an electro-optical/infrared (EO/IR) imaging camera and a Signals Intelligence (SIGINT) collection array. Each package weighs 1000 lbs., based upon an assumption the current 2000 lb. RQ-4 EO/IR and SIGINT payloads operating at similar altitudes can be reduced in weight over the next 20 years.[53] For the WSAs to employ SA-SDBs, they must also carry a targeting and communications package weighing 200 lbs. This package contains the electronics to compute bombing profiles and communicate with the JTAC, functions required for SA-SDB operations. Each SA-SDB requires 100 lbs. of payload capacity: 80 lbs. for the SA-SDB and 20 lbs. for the bomb rack/ejector. Table 1 summarizes the weights of the ISR packages, bomb assemblies, and SA-SDB munitions used in the variants discussed below.

ISR Package (each): (Interchangeable EO/IR and/or SIGINT)	1000 lbs.
SA-SDB Targeting and Communications Suite: (required for SA-SDB Ops)	200 lbs.
SA-SDB Munitions (each) (80lbs. + 20 lbs. for rack/launcher)	100 lbs.

Table 1: Payload Capacity Constraints

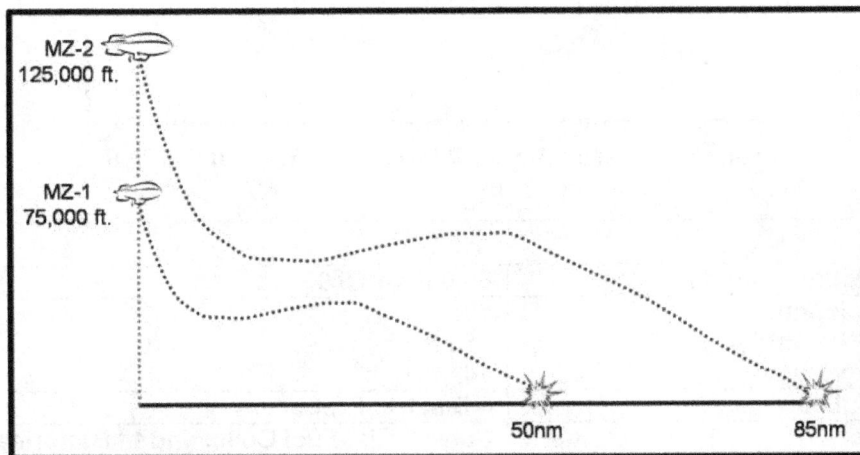

Figure 7: Drop Profiles of SA-SDBs from 75,000 ft. and 110,000 ft.

MZ-1: High Altitude WSA (65K – 85K feet altitude)

The SMDC HAA program is the basis for the MZ-1. Four electric propellers provide propulsion and a combination of solar cells, fuel cells, and lithium ion batteries powers both the propulsion systems and the payloads. With 4000 lbs. of payload available for ISR or munitions, commanders can select the best grouping of payload capabilities based upon mission demands. The MZ-1's ISR packages will have a view to the horizon line approximately 290 nm. away, providing ISR coverage over an area of 265,000 square nautical miles. The SA-SDBs will have a range of 50 nm. providing the MZ-1 capability to hit targets within a 7800 square nautical mile area. Figure 8 displays MZ-1 munitions and ISR ranges and Table 2 summarizes MZ-1 statistics.

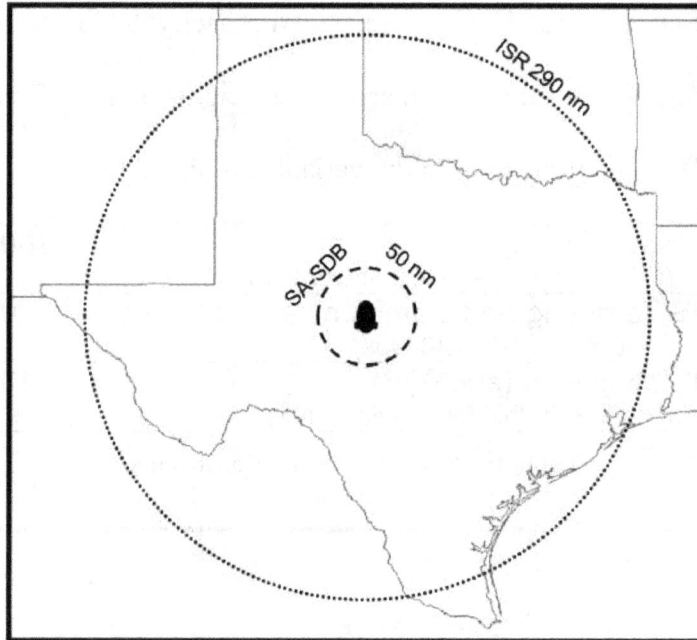

Figure 8: ISR and SA-SDB Ranges of the MZ-1 at 75,000 ft.

MZ-1	
Operating Altitude	65,000 to 75,000 ft.
Size: length	500 ft.
height	150 ft.
volume	147,000 m^3
Propulsion	4 Electric Propellers
Power	Solar Cells, Fuel Cells, and LI Batteries
Payload Capacity	4000 lbs.
# Munitions: No ISR Package	38
1 ISR Package	28
2 ISR Packages	18
Max Munitions Range	50 nm.
Max Munitions Time to Target	13 mins
ISR Range to Horizon (75K ft.)	290 nm.
Available Loiter Time	1 year

Table 2. Basic Operational Statistics of the MZ-1

MZ-2: Near-Space WSA (110K – 130K feet altitude)

The NASA Ultra Long Duration Balloon (ULDB) program is the basis for the MZ-2. The ULDB is an unsteered balloon for operations above 100,000 feet with a 6000-lbs. payload.[54] This paper assumes industry can develop an airship variant of the ULDB technology; however, the extra composite structural materials, propulsion system, and power elements will reduce its payload capacity to 2000 lbs.

Six electric propellers provide MZ-2 propulsion and a combination of solar cells, fuel cells, and lithium ion batteries power both the propulsion system and payloads. The MZ-2's ISR packages will have a longer view to the horizon of approximately 370 nm. providing ISR coverage over an area of 425,000 square nautical miles. The MZ-2's SA-SDBs will have a range of 85 nm. providing the capability to hit targets within a 22,500 square nautical mile area. Figure 9 displays the munitions and ISR ranges and Table 3 summarizes the statistics of the MZ-2.

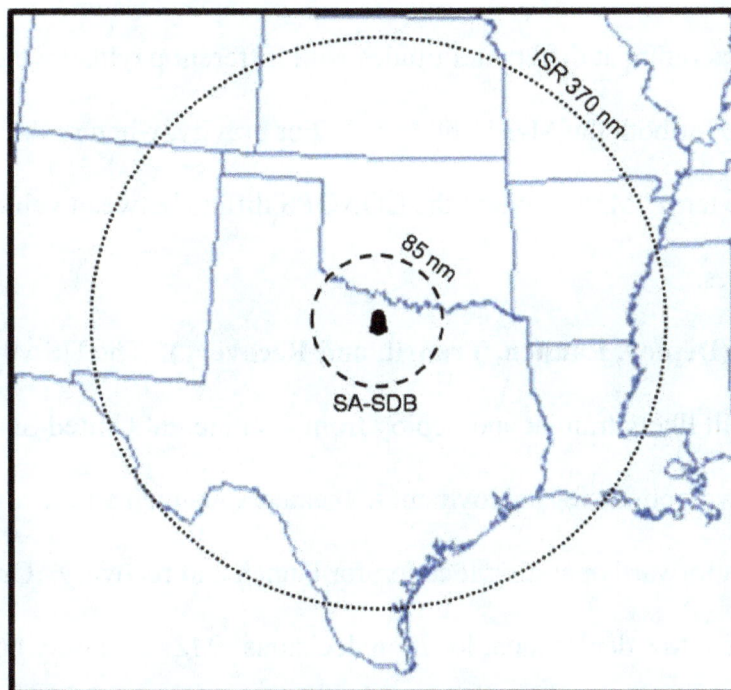

Figure 9: ISR and SA-SDB Ranges of the MZ-2 at 120,000ft.

MZ-2	
Operating Altitude	110,000 to 130,000 ft.
Size: length	1000 ft.
height	300 ft.
volume	632,000 m^3
Propulsion	6 Electric Propellers
Power	Solar Cells, Fuel Cells, and LI Batteries
Payload Capacity	2000 lbs.
# Munitions: No ISR Package	18
1 ISR Package	08
2 ISR Packages	00
Max Munitions Range	85 nm.
Max Munitions Time to Target	24 mins
ISR Range to Horizon (120K ft.)	370 nm.
Available Loiter Time	1 year

Table 3. Basic Operational Statistics of the MZ-2

MZ-1 and MZ-2 CONOPS

This section will define a top-level CONOPS for the deployment (launch, transit and recovery), employment (ISR and weapons release), and command and control (C2) for both variants. Although operating at different altitudes with different payload weights, the basic CONOPS is the same for both the MZ-1 and MZ-2. For brevity, when referring to both vehicles, the paper will use the term "MZs." When the CONOPS differs between vehicles, the paper will identify the differences.

Deployment (Deploy, Launch, Transit, and Recovery). The USAF squadrons operating the MZs will likely train at and deploy from Continental United States (CONUS) bases. The MZs will have two options for deployment to theater: 1) launch and recover from a CONUS base or 2) deploy to a forward operating location for launch and recovery. Operational need for the system should drive MZ deployment location decisions. MZs will only be able to transit at 50 knots and thus require up to 10 days to reach their station from a CONUS location. Transit will occur at operational altitudes to reduce stresses on the vehicles; however, the WSA can exploit winds at different altitudes to speed its transit. The MZs will require approximately two hours to reach operational altitude and require approximately ten hours to descend from their operating altitude.[55] If a mission anticipates using a large number of munitions over a short time, a deployed location closer to the operating area will maximize MZ time on station versus in transit. A deployed MZ unit will travel with maintenance tools, spare parts, SA-SDB munitions, ISR packages, replacement helium supplies, portable hangars, and docking systems for recovering, maintaining, and protecting the MZs at deployed locations. Due to its immense size, three-times that of the MZ-1, the MZ-2 will require a far larger deployment footprint than the MZ-1. The MZ-2's size will also result in more restrictive weather requirements when launching and recovering.

The USAF should select MZ home bases for location and favorability of weather. Since the MZs will be vulnerable to weather during ascent, descent, launch, and recovery operations, a favorable weather location will minimize impacts to operations and training. Coastal bases will also reduce the transit times to overseas locations by eliminating travel time over the US.

Operations. Operators will command MZs remotely via SATCOM from a permanent Ground Control System (GCS) facility at the home base. A deployable, containerized GCS can also forward deploy and co-locate with the launch/recovery base, a CAOC, or the JFC's intelligence center. Many of the MZs' day-to-day functions will be automated. Using GPS and onboard systems, the MZ will be able to transit independently along a programmed course to an operating position or recovery base. The MZs will provide their GCS with constant system status via SATCOM to include position, geometry, airspeed, and temperature as well as status on the helium envelope, power, propulsion, and payloads. The data stream from the ISR packages will also be sent via SATCOM. Since many of the MZ functions will be automated and the MZs will move at a slow pace, a single pair of operators at a GCS can operate several MZs.[56]

The ISR packages aboard the MZs will be commanded by the GCS or independently by a separate workstation located in an operations or intelligence collection center. Depending upon the number of MZs available in theater, some MZ missions will be dedicated only to ISR with other missions dedicated only to SA-SDBs. The CAOC will reallocate MZ missions as the JFC's requirements balance changes between ISR and SA-SDB weapons effects.

Three methods will be available for SA-SDB weapon programming and release. The first is the GCS directly commanding SA-SDB release. The second is via a separate MZ weapons workstation located in a CAOC TST cell. The third method is an MZ Airship Handheld Control (MZAHC). The MZAHC is a handheld computer for communication with the MZ via digital data and voice link over HF radio. JTACs and other specially trained personnel will carry and operate MZAHCs. The GCS can grant an MZAHC operator release functionality for an SA-SDB. An MZAHC will facilitate programming of target coordinates or targeting laser frequencies into an SA-SDB. Prior to release, the MZ will confirm all targeting data by sending a verification message back to the JTAC for confirmation. This verification will also include an estimated time-to-impact of the SA-SDB once released. The JTAC will also use the MZAHC as his communications method to request weapons release from the GCS MZ operator.

Command and Control. When deployed, a COMAFFOR will exercise Operational Control (OPCON) of an MZ. The CFACC will hold TACON of the MZ and daily tasking for the airship, ISR packages and munitions will be published in the ATO. The ATO will assign one of four options for each SA-SDB on an MZ. The first option is a pre-designated target with a specified time-on-target for the munition. The second option is to a specific Air Support Operations Center (ASOC) for JTAC-supported CAS missions. The ATO will specify whether the SA-SDB will be controlled by a JTAC's MZAHC or by the GCS in coordination with a JTAC. The third option is TST missions. The CAOC TST cell will have the munition available for release via their WSA workstation when a TST is located. The fourth option is a reserve

mode. The ATO may hold SA-SDBs in reserve for upcoming operations in future ATOs. The

ATO will also set "ditch" targets for each SA-SDB. The ditch targets will be for situations

where enemy attack or adverse weather disables the MZ and it is unrecoverable. Operators

would then command the MZ to drop its SA-SDBs on ditch targets so the munitions do not fall

with the MZ into enemy, friendly, or neutral territory. A "ditch" target could be an unpopulated

area such as a desert or lake, or an enemy target, such as an airfield or army post.

The CFACC can utilize the MZs for a variety of missions. The three most common

missions are ISR, CAS, and TST. The next section will apply the MZ variants to these missions

using two scenarios.

WSAs in Action

This section develops two scenarios and applies both WSA variants to them. The first scenario is the Low Intensity Conflict (LIC) stability operations of Iraq and Afghanistan over the past six years. It involves a negligible surface-to-air and air-to-air threat and assumes a permissive environment for the launch and recovery of WSAs in theater. The second scenario involves US defense of Taiwan against invasion by the People's Republic of China, a near-peer competitor. The scenario involves a significant surface-to-air and air-to-air threat in which the options for launch and recovery of an airship will be significantly more constrained.

Low Intensity Conflict: Iraq and Afghanistan

Scenario Description and Development. Several inherent differences exist between air operations in conventional conflict and low intensity conflict (LIC). Lt Col Phil Haun, a LIC expert from OIF wrote, "In LIC there are no enemy aircraft to engage, no enemy air defenses to attack, no state headquarters to surgically strike, and no fielded forces to interdict. Airpower still has a critical role to play, but it typically supports the occupying ground forces. These missions include tactical airlift; intelligence, surveillance, and reconnaissance; and LIC CAS."[57] The USAF conducts CAS and TST missions against an obscure adversary in a low-threat environment like Iraq or Afghanistan. Targets for aircraft are few and far-between, a sign of successful progress in a LIC, but a frustrating and costly endeavor for manned aircraft due to the need for constantly orbiting, and rarely utilized, fighter aircraft.[58] Statistics from Iraq show that sorties employing weapons account for only four percent of all fighter missions flown; however, lethal fires from the air make a critical difference when friendly troops are in contact with enemy forces or critical targets require destruction.[59] Even though rarely used, the persistent availability of air firepower is critical to progress in a LIC.

Close Air Support missions in Iraq and Afghanistan typically involve Troops in Contact (TIC). During TIC, ground commanders will sometimes request air support to destroy insurgent positions in support of friendly ground force objectives.[60] However, not all TIC missions are ideal for CAS. Commanders may choose to avoid engagement from the air if a TIC is in an urban environment or other location prone to fratricide or collateral damage. Such engagements may produce tactical success, but damage strategic progress by upsetting the local populace.[61]

Destruction of Time-Sensitive Targets (TSTs) also plays a critical role in a LIC like Iraq and Afghanistan. Insurgents practice "shoot-and-scoot" tactics: quickly setting up to engage US forces with a mortar or other device, executing the attack, and then quickly dismantling the weapon and moving on.[62] If timely identified, evaluated, and engaged, these insurgents present fleeting targets ideal for destruction by airpower. Another TST situation successfully executed in Iraq with high payoff was the attack on "al Qaeda in Iraq" leader, Abu Musab al-Zarqawi. US troops monitored a suspected safe house for several weeks and called in an air strike after confirming Zarqawi's presence at the residence.[63] Persistently available air support enabled the timely destruction of this critical target.

Another aspect of airpower that has been useful in Iraq and Afghanistan is show-of-force. In certain high-collateral-damage situations such as TIC in urban environments, commanders do not want aircraft to fire on the enemy. US aircraft instead exercise show-of force with loud, low-level passes over targets to frequently scare off the enemy forces. Show-of-force can also be successful in preventing crowds from complicating ground force efforts by encouraging large groups to disperse.[64] Visible airpower, such as CAS aircraft escorting convoys or troops on patrol, has also acted as a deterrent to insurgent attacks on Coalition forces.[65]

WSA Missions for Low Intensity Conflict: Iraq and Afghanistan. In this LIC scenario, the MZs perform three primary missions: ISR, CAS, and TST. LIC is intelligence intensive, requiring detailed information to identify adversaries amongst the population; therefore, the ISR provided by the persistent MZs is in high demand.[66] JTACs have SA-SDB munitions available to them when ground commanders request CAS for engagement of targets during TIC or other situations involving enemy insurgents. The CAOC TST cell also has SA-SDBs available for use when they identify TSTs.

Use of the MZs in the conflict could reduce the number of Coalition fighter aircraft needed to pursue the LIC air needs. This would decrease the footprint of US forces in-country as well as in neighboring countries. The theater will require fewer expensive fighter CAS sorties and their associated support and personnel, thus providing lower operational cost and a decreased US regional presence. Because the US has basing agreements with countries in the region and the weather is favorable for MZ operations, the USAF has forward deployed MZs to the Persian Gulf region. This forward deployment minimizes off-station time when an MZ must return to base to reload SA-SDBs or change out ISR packages.

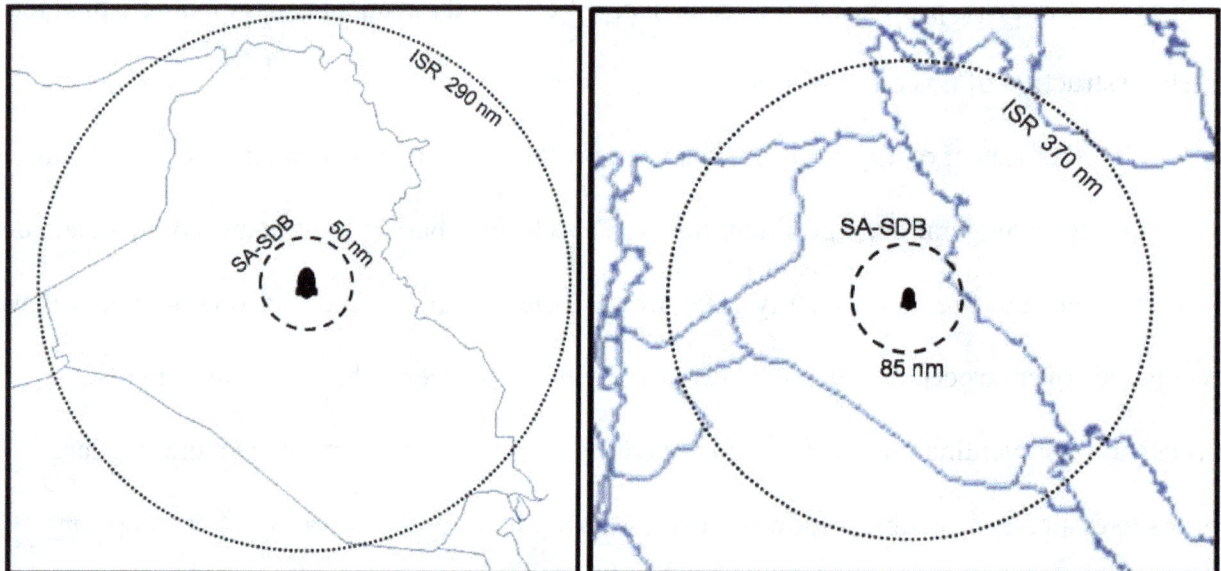

Figure 10: MZ-1 and MZ-2 Ranges over Iraq

Use of MZ-1 in Low Intensity Conflict: Iraq and Afghanistan. The MZ-1 performs well in the ISR and TST missions. The CAOC and ground commanders have made good use of the persistent imagery and SIGINT available from above. The CAOC TST Cell has also destroyed several short-notice insurgent targets using their assigned SA-SDBs. In several cases, insurgent mortar teams have made the mistake of returning to a location under surveillance by the MZ-1. SA-SDBs destroyed them in an average of six minutes from detection.

The MZ-1 has received lukewarm response from ground commanders for CAS support. Due to concerns with the SA-SDB's time-to-target, there are few situations where the target's lack of movement allows the commander and his supporting JTAC to await the arrival of the SA-SDB up to 13 minutes later. Because of this limitation, commanders have rarely used SA-SDBs in urban engagements. However, in situations where the targets are rural or contained to a single location, the MZ-1's persistence has enhanced the ground effort.

Use of MZ-2 in Low Intensity Conflict: Iraq and Afghanistan. The MZ-2 has also been useful in ISR situations described above by providing ISR over an even broader area. Though SA-SDBs have a longer range from an MZ-2, they have received less use because of the extensive time-to-target. Additionally, the MZ-2s low munitions load and longer turn-time to replenish munitions has created difficulties. The CAOC has now assigned the MZ-2s primarily to ISR duties with their limited munitions only used for the highest priority TSTs. Though they have less utility in a LIC, the MZ-2s provide more utility in a conventional scenario.

Conventional Conflict with a Near Peer: PRC vs. Taiwan

Ever since retreating from mainland China to Taiwan in 1949, the Republic of China (ROC) has experienced tenuous relations with the People's Republic of China (PRC). The relationship between the two entities is still complicated with the PRC desiring to one-day re-unite with their "renegade province." The PRC may attempt this reunification by friendly, coercive, or military means. The US desire to promote democracy, encapsulated in national strategy and the Taiwan Relations Act, encourages any reunification between the two sides be accomplished peacefully. To promote peaceful resolution, the US has supported Taiwanese defense with arm sales and expressed a willingness to aid Taiwan's defense if attacked.[67] On the other side of the Taiwan Strait, the PRC People's Liberation Army (PLA) has been strengthening its military capabilities, increasing the threat to Taiwan.[68] If the US utilizes military force in defense of Taiwan, it will face several challenges from the PLA.

In order to thwart a PRC invasion, US and Taiwan airpower will need to assist defensive efforts by degrading PRC military capabilities on the mainland. This effort will also include action over Taiwan should PRC forces successfully gain access to the island.[69] Commanders will call upon airpower to destroy ballistic missile sites, seaports, airfields, and PRC ground units. Additionally, if PRC forces land on Taiwan, the JFC will apportion airpower to provide CAS in support of Taiwan and US ground forces defending the island.[70] Several PRC capabilities exist to thwart US and Taiwanese airpower from these objectives.

The PRC possesses large numbers of fighter aircraft. Amongst the most capable are the J-10 and the J-11. The J-10 is a PRC produced multi-role fighter capable of Mach 1.85 speeds and a ceiling of almost 60,000 ft.[71] The J-11 is a PRC version of the Russian Su-27 and Su-30 fighter. It is capable of Mach 2.3 speeds and has a service ceiling of 57,000 ft.[72]

To thwart adversary air activity, the PRC has also heavily invested in its integrated air defense system (IADS). The current system includes PRC variants of the Russian SA-10 and SA-12 surface to air missiles (SAMs). Each system is capable of hitting targets out to 125 miles and as high as 100,000 ft.[73] In the future, the PRC will likely add a directed energy (DE) capability to their IADS arsenal. A DE weapon will likely be capable of creating destructive effects out to 50nm and up to 65,000 ft. altitude. It could produce degrading effects out to 150 nm. and above 150,000 ft. altitudes.[74]

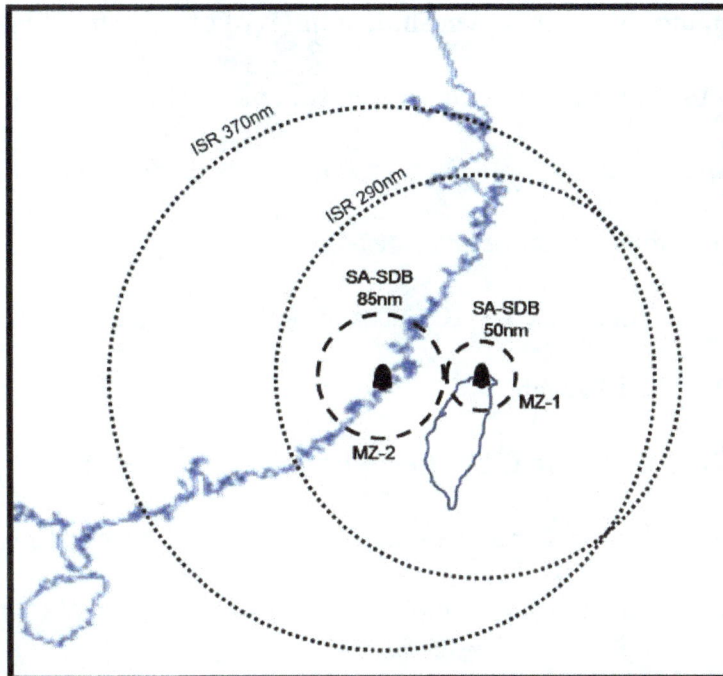

Figure 11: MZ-1 and MZ-2 Ranges over the Taiwan Strait

WSA Missions for Conventional Conflict with a Near Peer: PRC vs. Taiwan. When

indications and warnings tipped off the Commander in Chief to an imminent PRC invasion of

Taiwan, multiple MZs launched within a day and were on station within six days. The initial

loiter positions were set so the MZs did not violate PRC airspace. With their long ISR range, the

MZs collected on PRC intentions and order of battle. When the PRC initiated hostilities by

launching several ballistic missiles into Taiwan, the CFACC immediately tasked several MZs to

drop munitions on known launchers. Unfortunately, the PRC quickly learned of this

vulnerability and all mobile launchers took up the tactic of moving immediately after missile

launch.

Quick expenditure of SA-SDBs has become an issue. Because of the vast expanse of the

Pacific Ocean, both MZ-1s and MZ-2s must travel considerable distances to friendly bases.

When fielded, the USAF established forward operating facilities for the MZs on Okinawa, Guam,

and Hawaii. A majority of the MZs supporting the effort operates from these bases; however, the MZs require minimum roundtrip replenishment times of one day to Okinawa, three days to Guam, and eight days to Hawaii. Areas on Taiwan suitable for MZ basing are under too high a threat from PRC missile and air attacks. On many occasions, MZs spend more time in transit to replenish their munitions than on station over theater.

As the conflict continued, the CFACC tasked both MZ-1s and MZ-2s to ISR, CAS, and TST duties over the PRC and Taiwan.

Use of MZ-1 in Conventional Conflict with a Near Peer: PRC vs. Taiwan. The MZ-1s were susceptible to PRC actions. Early in the conflict, MZ-1s successfully dropped munitions on TSTs such as SAM sites and missile launchers. After realizing the WSA threat, the PRC placed an emphasis on identifying and attacking MZ-1s. Using advanced optics and radar tracking of SA-SDB drops, the PRC determined general locations of the MZ-1s. SA-10s and J-10 fighters successfully attacked some of the MZ-1s. Proximity explosions and fragmentation did not immediately disable the MZ-1s; however, due to the lack of nearby bases, most had to ditch in the Pacific Ocean as they slowly lost maneuvering capability due to helium envelope deflation. As they departed PRC airspace, the damaged MZ-1s struck final blows by dropping their remaining SA-SDB munitions on known airfields and PLA facilities.

The CFACC assigned several MZ-1s to stations over Taiwan, but they quickly became susceptible to SAM fire from the mainland. Ground commanders also never fully used MZ-1s as CAS weapons. When PRC forces invaded, the situation was too fluid for the JTACs to hit moving targets with SA-SDBs. Additionally, US and Taiwanese CAS aircraft saturated the airspace over ground forces and the CAOC was unable to clear it for SA-SDB engagement.

Use of MZ-2 in Conventional Conflict with a Near Peer: PRC vs. Taiwan. The MZ-2s enjoyed success as both an ISR and TST platform over the PRC mainland. If the PRC was able to initially determine their location, they never successfully engaged the MZ-2s due to their high altitude. Early in the conflict, the MZ-2s were very successful in destroying SAM sites as soon as the MZ-2s SIGINT package detected them. The CFACC assigned all SA-SDBs to TST SAM targets in support of the high-priority JFC Air Superiority objective. Unfortunately, the MZ-2s quickly depleted themselves of munitions in the first few days of the conflict. Because of their value as ISR platforms, the JFC was unwilling to release MZ-2s back to their bases for SA-SDB replenishment.

Later in the conflict, the MZ-2s also began to lose utility due to PRC actions. Using DE weapons, the PRC was able to damage the sensitive ISR packages and in one case, even burned a puncture causing an MZ-2 to ditch in the Pacific Ocean. The MZ-2s reliance on SATCOM also quickly became a vulnerability. Since the MZ-2s were transmitting from over PRC territory, the MZ-2 communication links to satellites above were constantly jammed and ISR data lost. By modifying tactics and commanding constant SATCOM frequency changes, operators were occasionally able to overcome the PRC jamming.

Scenarios Summary

The MZ variants assisted the USAF and JFC towards achieving military objectives in both scenarios. Both MZs provided long-term persistence to observe targets and drop available munitions when tasked. The MZs performed adequately on their two assigned weapons missions: CAS and TST. Several missions performed by current CAS aircraft will be difficult for the MZs. Since the MZs sit quietly at high altitudes, they cannot execute show-of-force missions similar to fighter aircraft. Although the ISR packages on the MZs will have good resolution, an MZ operator sitting half a world away will not have the same CAS mission situational awareness as a pilot close to the target. This lack of situational awareness can hinder support to ground search missions as well as danger-close and fluid CAS missions. The MZs performed well on TST missions. Their persistence provided the ability to hit high-value targets, such as insurgents and missile launchers, as soon as they were detected. Several tenets made these missions possible.

The WSA has adequate payload capacity, excellent munitions range, and excellent survivability with some recovery issues. The payload capacity of the MZ-1 is suitable for weapons missions; however, the limited payload of the MZ-2 limits it to only high-priority targets. Both systems provide excellent munitions range with the MZ-2 being able to cover an area over 170 nm. in diameter. The MZ-1 proved to be survivable in certain environments with the MZ-2 able to avoid most threats. A main strength of the MZ-1 over the MZ-2 is recoverability at a deployment base. Due to its size, the MZ-2 is far more limited by weather.

The MZ-1 and MZ-2 provided unique and persistent ISR capabilities in the two scenarios. Though there are limitations to their utility, they definitely aided the JFC and CFACC towards achieving their objectives. Their persistence and survivability are one of many assets discussed in the summary.

Summary

WSAs demonstrate several tenets providing a strong and unique capability to joint force commanders. The multiple-month persistence of WSAs will enhance ISR and available weapons effects in an area of operation. Few commanders ever state they have too many ISR assets on station and the ability to drop a weapon on a moment's notice within a 170-nm. diameter is intriguing. Their mission will be greatly supported by additional long-range ISR assets overhead as well as munitions that can hit targets from 50 to 85 nm. away. WSAs, due to their high altitude, low observability, and limited impact from damage, are inherently survivable over a war zone. The airship platform can also loiter, waiting for a CAS or TST mission at a fraction of the operational cost of today's fighter aircraft. The combination of these factors, persistence, range, survivability, and cost-effectiveness, make them strong platforms for CAS and TST of stationary targets. These capabilities also have significant restrictions.

WSAs have several limitations that will impede their effectiveness as high altitude CAS and TST platforms. Due to low payload weight capacity, the number of munitions carried by each WSA is limited. Commanders may face situations where the WSA has expended all its munitions long before it needs to leave its station. Even though the weapons capability is gone, a WSA can still perform its ISR mission. Due to slow transit, WSAs also require significant travel time to their recovery bases. This travel time can be compounded by weather at the recovery bases as well as the distance from the operating location to those bases. The high altitude, persistence and range of WSAs also create significant limitations for CAS and TST missions. Since the munitions on a WSA require significant time to reach targets at long range, JTACs or TST Cells may be hesitant to request munitions from WSAs in rapidly changing situations where targets are likely to move. In certain CAS situations, a WSA also provides far less flexibility than CAS airplanes because the WSA operator has less situational awareness. The WSA also cannot perform show-of-force flyovers as an intermediate step to resolving a situation prior to weapons release. Highly dependent upon SATCOM for operations, WSAs may be vulnerable to jamming by enemy forces. These limitations contribute to some top-level WSA issues.

Two major issues must be resolved if WSAs are to become a viable future weapons platform. The first issue is competition between the WSA's ISR and firepower missions. WSAs require significant time to replenish munitions and return to station. This will likely create situations where commanders are reluctant to release a "Winchester" WSA (out of munitions) from its ISR mission for munitions replenishment. The second issue is the feasibility of high-altitude stratospheric airships. Though currently strong on paper, viable stratospheric airships carrying the payload weights discussed in this paper have yet to fly. Several technical issues must be resolved before WSAs become a reality. The main issue is airship size versus weight tradeoffs. A WSA large enough to carry a significant payload may be too large for effective ground operation or maneuverability in the air. Other technical issues include viable lightweight power systems and effective lightweight propulsion systems.

Capabilities	Limitations	Issues
- Persistence	- Limited Munitions	- Airship Viability
- ISR Range	- Replenishment Time	- ISR/Fires Mission Conflict
- Munitions Range	- CAS Situational Awareness	
- Survivability	- No "Show-of-Force"	
- Cost Effectiveness	- SATCOM Jamming	

Table 4. Weaponized Stratospheric Airship Capabilities, Limitations, and Issues

Other Considerations

This paper provides only a cursory glimpse at the issues related to the feasibility of

WSAs as effective weapons platforms. Several other avenues could be explored. The first is low altitude airships. While more vulnerable to enemy action, airships at lower altitude have an exponentially larger payload capacity than high altitude airships allowing the carriage of more munitions. One example is the DARPA Walrus concept, an airship operating below 10,000 ft. with a payload capacity of over 500 tons.[75] An option to overcome the time-of-flight issue is miniature cruise missiles such as the SMACM. Although they would likely weigh more than the SA-SDBs, a thrusted munition may have longer range at faster speed. An option to better support moving target situations is to add capability to the SA-SDBs for retasking in flight. Finally, if the airships are unable to maintain station-keeping due to winds or other factors, the USAF could explore using multiple balloons or airships and allowing them to float one-by-one over the target area, collecting ISR and dropping munitions when required.[76] These considerations are for naught if the Air Force is unwilling to accept the WSA concept due to institutional biases.

Institutional Acceptability

> *You never go to an air show to go watch a balloon performance. They don't put on a very good acrobatic show and it's just not very cool.*

—General John P. Jumper

A major issue for WSAs not previously covered in detail involves the institutional acceptability of "old" technology. In 2005, while serving as USAF Chief of Staff, General Jumper made the above remarks to the Heritage Foundation as way of explaining the perceptual hurdles lighter-than-air craft will face assimilating into the Air Force.[77] WSAs are a vastly different philosophy from the aircraft and space assets dominating today's Air Force. Though they utilize advanced technology, SAs are reminiscent of the Hindenburg and other airships of the early 20th century. This creates a perception of old technology--flying in the face of the Air

Force's institutional emphasis on new technology. SAs are also a slow platform, contrary to the perception of today's "high-speed, low-drag" Air Force. If SAs gain traction, they will also likely pose a budget threat to existing programs because their effects, ISR and firepower, are currently provided by airplanes and satellites.[78] To successfully win approval, WSAs must overcome the old and slow biases, as well as provide capabilities and effects significantly better than, different from, or cheaper than the capabilities of existing assets.

Conclusion

Although weaponized stratospheric airships possess limited combat utility, the Air Force should pursue the technology as a persistent means of providing close air support and timely destruction of time-sensitive targets. Their persistence, ISR and munitions range, survivability, and cost-effectiveness provide an intriguing and unique capability for future protection of US national interests. No system is ever able to accomplish all missions on its own, thus the WSA strongly compliments existing ISR, CAS, and TST platforms. Recent USAF funding commitments to the DARPA ISIS program show the DOD is moving in the right direction towards implementing stratospheric airship concepts. The USAF and DOD should also continue to pursue small munitions technology for airplanes and UAVs with thoughts toward implementing them on future stratospheric airships. Stratospheric airship programs, small munitions programs, and tactics developers should all be collaborating on means of creating and exploiting the merging of the technologies to ensure the USAF fully takes advantage the WSA concept. These actions will put the weaponized stratospheric airship on a path to success-- providing persistent, survivable, and cost effective ISR and munitions effects to future warfighters.

1 A "nine-line" is the standard target data exchanged between a JTAC and the operator of the CAS attack aircraft as called out in JP 3-09.3, "Joint TTPs for Close Air Support."
2 Amol, "Floating a New Idea."
3 Sanswire, "Unmanned Airship Solutions."
4 Lockheed Martin, "High Altitude Airship."
5 Military and Aerospace Aeronautics Magazine, "Northrop Grumman to Build."
6 JP 3-09.3, "Joint TTPs for Close Air Support."
7 Stephens, "Near Space," 37.
8 Ibid, 37.
9 Tomme, "The Paradigm Shift," 43.
10 Ibid, 43.
11 USAF Fact Sheet, "TARS."
12 Mayer, "Lighter-Than-Air Systems," 30.
13 Lockheed Martin. "High Altitude Airship (HAA)."
14 Schecter, "Airships on the Rise," 30.
15 "Northrop Grumman to Build."
16 Schecter, "Airships on the Rise," 30.
17 Cathay, "Development of the NASA," 13.
18 Kondrack, "High Altitude Airship Station-Keeping," 25.
19 Chu, "A Novel Concept," 1.
20 Sanswire, "Unmanned Airship Solutions." 2.
21 Smith, "Applications of Scientific Ballooning," 3.
22 Schecter, "Airships on the Rise," 30.
23 Kondrack, "High Altitude Airship Station-Keeping," 99.
24 Schecter, "Airships on the Rise," 30.
25 Torbinson, "Lockheed Martin Makes."
26 US DOI, "Materials Summary," 78.
27 GlobalSecurity.org, "S-300PMU."
28 Schecter, "Airships on the Rise," 30.
29 Tomme, "The Paradigm Shift," 44.
30 Ibid, 13.
31 Vogt, "Performance Capability of a Damaged," 63.
32 Tomme, "The Paradigm Shift," 13.
33 Schecter, "Airships on the Rise," 30.
34 Cathay, "Development of the NASA," 13.
35 Blackington, "Near Space Maneuvering Vehicle," 3.
36 Sanswire, "Unmanned Airship Solutions," 2.
37 ACC, "SDB Fact Sheet."
38 Vogel, Email Dated 7 Nov 2008.
39 Boeing, "Small Diameter Bomb Increment I."
40 Boeing, "Small Diameter Bomb Increment II."
41 Boeing, "SDB Focused Lethality Munition."
42 Vogel, Email Dated 7 Nov 2008.
43 Borden, "Viper Strike."
44 Lockheed Martin, "SMACM."
45 Pirnie, "Beyond Close Air Support," 167.
46 Brown, "JTAC: MOA vs. MTTP," 18.
47 Pirnie, "Beyond Close Air Support," 117.
48 Absolute-Astronomy.com, "Cumulonimbus Cloud."
49 JP-3.60, "Joint Targeting," I-5.
50 Ingram, "Joint Targeting for," 28.
51 Haffa, "Command and Control Arrangements," 32.
52 AFI-16-401, "Designation and Naming, " 13.
53 USAF Factsheet, "RQ-4 Global Hawk."
54 Cathay, "Development of the NASA," 13.

[55] Jamison, "High Altitude Airships," 33.
[56] Tomme, "The Paradigm Shift," 34.
[57] Haun, "The Nature of Close Air Support," 107.
[58] Ibid, 107.
[59] Ibid, 108.
[60] Belote, "Counterinsurgency Airpower," 57.
[61] Haun, "The Nature of Close Air Support," 108.
[62] Belote, "Counterinsurgency Airpower," 57.
[63] "Iraq Terrorist Leader Zarqawi 'Eliminated.'"
[64] Belote, "Counterinsurgency Airpower," 58.
[65] Haun, "The Nature of Close Air Support," 108.
[66] Ibid, 107.
[67] US Department of State, "Background Notes: Taiwan."
[68] Fisher, "Deterring a Chinese Attack."
[69] Ibid.
[70] Ibid.
[71] GlobalSecurity.org, "J-10."
[72] GlobalSecurity.org, "J-11/Su-30."
[73] GlobalSecurity.org, "S-300PMU."
[74] Thill, "Penetrating the Ion Curtain," 19.
[75] Gordon, "Back to the Future," 53.
[76] Kondrack, "High Altitude Airship Station-Keeping," 102.
[77] Stephens, "Near Space," 38.
[78] Tomme, "The Paradigm Shift," 16.

Appendix A: Stratospheric Airship Environmental Scan Details

This appendix provides additional environmental scan details on the requirements of a weaponized stratospheric airship (WSA) and the capabilities of current and future technologies.

SA Requirements

This section will discuss current and forecasted technology for stratospheric airships and small munitions to develop and range of potential capabilities for weaponized stratospheric airships (WSAs) in the 2030 timeframe.

The primary concern is the operational environment between 60,000 and 130,000 feet (20 to 40 km). The air in this range is far less dense than at normal operational altitudes for aircraft, and significantly less dense than sea level. Air density at 100,000 feet is one percent of air density at Sea Level and two percent of air density at 18,000 feet.[79] This lack of air density will require a large volume of buoyant gas in order for the airship to float at those altitudes. While the air density makes the region difficult for operations, winds are more favorable. Winds at 60,000 to 100,000 feet are above both storms and the jet stream so they tend to range from only 20 mph to 40 mph. Even at 130,000 feet, current models shows winds to average less than 50 mph.[80] With proper steering and propulsion, a stratospheric airship should be able to maneuver or maintain a stationary position at these altitudes.

In this environment, the primary requirements of a WSA are to loiter over an area of interest for an extended period and drop precision munitions when commanded. This section will detail the following stratospheric airship capabilities required to make the WSA a reality: maneuver/station-keeping, payload capacity, survivability, and sustainability. Each section will discuss why the capability is vital to the WSA, the current and future outlook for these capabilities, and issues associated with the capability.

Maneuver and Station-Keeping

Maneuver/station-keeping is ability of the WSA to move from its launch point to its loitering position, maintain that loitering position, and then move to a point of recovery, which may be the same point from which the WSA launched. Maneuver/station-keeping is primarily a function of three factors: structure, propulsion, and power. The structure provides the lift but due to size, the structure also impedes maneuverability. The propulsion system provides the thrust to enable maneuverability. The power system provides a constant source of energy for the propulsion system as well as the payload. Each factor will be covered in detail below.

Structure. Several factors must be taken into account for the structure of a stratospheric airship. It must be aerodynamic so it can maneuver within the air without significant resistance and maintain stability for the payload. Through the air at operational altitude is far less dense than at sea level, aerodynamics must still be taken into account because the less dense air will also reduce the effectiveness of the airship's propulsion system.[81] The structure must also be able to provide buoyancy by containing the gas that provides the airship's life. A system must also be in place for buoyancy control, allowing the system to adjust its altitude. This can be done with changes in pressurization or ballast with the goal of designing a system that uses the least possible energy and does not cause the loss of lifting gas.[82] Finally, the airship must be manufactured out of a lightweight material that can maintain internal pressure while also withstanding the temperature extremes of being exposed to solar radiation in the daytime and extreme cold at night.[83] Two examples of airship types provide examples of these concepts.

The Lockheed Martin High Altitude Airship (HAA), slated for flight in late 2009, is a large airship with the traditional airship shape of historical airships such as the Hindenburg and the Goodyear blimp. It uses a combination of its propulsion system and fins to control movement and promote stability. Buoyancy is provided by helium and controlled by use of multiple internal compartments inside of a high-strength fabric to minimize airship weight.[84] In order to provide enough buoyancy for its payload at its planed operational altitude of 65,000 ft., the HAA is 150 feet tall, 500 feet long, and contains as much helium as 25 Goodyear blimps.[85]

A second structure type is a V-shaped design. Before it was inactivated in 2006, the USAF Space Battlelab was pursuing its Near Space Maneuvering Vehicle (NSMV) project. The NSMV intended to build a prototype airship consisting of Mylar fabric over a composite frame for rigidity. The airship shape would consist of two tubes extending from the front point in a v shape. The NMSV's V-shape would provide both stability and reduce drag when maneuvering with its electric motors.[86] The project intended to launch the prototype NMSV with a 100-pound communications payload, maneuver 200 nm. downrange, loiter for 120 hours, and then return to its launch site.[87]

A third structure type with potential for higher altitudes above 100,000 feet is a steered free-floater. Dr. Edward Tomme, near-space expert, calls this concept the "sailboats of near-space."[88] This structure is less like an airship because it has no control surfaces or propulsion system. The free-floater would maintain its position by extending a 15 km cable with a wing and rudder down into the air at lower altitudes. The differing winds between the altitudes allow the wing and rudder to effective steer the balloon at higher altitude in order to maintain position.[89]

The largest issue associated with airship structure is creating a structure large enough, strong enough, and light enough to provide both a significant payload capacity while also minimizing the power and propulsion required to maneuver the WSA. Results from the testing of the HAA in late 2009 will help determine a clear path ahead for the best structure of a future WSA.

Propulsion. The WSA's propulsion mechanism is the second factor for maneuverability and station-keeping. The propulsion system must provide enough thrust to propel the mass of the WSA against the drag of the standard 20 – 45 mph winds and occasionally greater gusts at operating altitude.[90] This is necessary to keep a stationary position over the WSA's area of interest. The propulsion system must also provide enough thrust to maneuver the WSA from its launch point to its station at altitude and then later return to a recovery point.

Electric propellers provide the best option for WSA propulsion. In 2005, Captain Eric Moomey wrote a thesis for the Air Force Institute of technology (AFIT) on the feasibility of a loitering near-space vehicle. His paper included an analysis of propulsion types and strongly concluded that an electric propeller is best mechanism. Other options included hydrazine thrusters, jet engines, and gas turbines. Because the other options require significant amounts of heavy fuel and/or oxidizer and electric propeller technology has already been proven at high altitudes on vehicles like the NASA Helios, the electric propeller is definitely the correct technology for WSAs.[91]

Another option for a WSA that would negate the need for a propulsion system would be a tether. A tether would be anchored to the earth with a winch or similar apparatus. The WSA could be raised to altitude and later lowered for re-arming and maintenance. The tether would also keep the WSA within a controlled distance around the anchor point. With a tether, the WSA would no longer be an "airship," but instead be an "aerostat." However, a tether is not a good option for the WSA. No current lightweight material exists that would have the strength over 20 km of tether to hold the WSA in position. The weight of the tether would also significantly reduce the payload capacity of the WSA. In the future, carbon nanotube wires or other materials may overcome this limitation. A tether would also create airspace issues for aircraft operating in beneath the WSA. Finally, the tether would negate WSA flexibility because the WSA could only be tethered in secure territory and could not maneuver far over enemy territory. [92]

Power. The WSA power system must produce enough electricity not only to power the electric propellers of the airship, but also the payload. The payload would require energy for the ISR package, flight computers, actuators on control surfaces, batteries in the munitions, and any environmental systems for sensitive components in the payload or munitions. Likely power sources include solar arrays, fuel cells and lithium ion batteries.

Solar cells would provide a regenerating power capability that could keep the WSA on station for a long period. However, due to darkness, the WSA would also require batteries to store electricity during daylight for use during periods of darkness. Factors such as temperature, propulsion needs, structural drag, and the efficiencies of the solar cells and batteries determine the final capabilities.[93] Solar cells and batteries also add significant weight to the airship robbing payload capability.

A second technology is fuel cells. Fuel cells have been used in manned spacecraft, cars, and aircraft as a lightweight source of power produced by a catalytic reaction between a reactant and oxidizer, typically hydrogen and oxygen. Since hydrogen fuel cells consume their reactant over time, they can only work for a specified length of time and require more weight the longer they are to work. Additionally, hydrogen fuel cells add significant logistical footprint to the WSA because of the storage and handling capabilities required for liquid hydrogen.[94]

The Lockheed Martin HAA makes use of both technologies for power proving the initial feasibility of both sources.[95] Due to their regenerative capability, solar cells are best for missions beyond 30 days while with a small fuel supply, fuel cells are best for shorter duration missions.[96] Solar Cell and Fuel Cell technologies have also been steadily improving over the past 5 years and will likely continue to improve.[97] At some point prior to 2030, both should be mature enough to provide the needed power for the WSA.

Maneuver/Station-Keeping Issues. The primary issue of current airship technology is the balance between airship structure size, payload, and power requirements to effectively keep a stratospheric airship carrying a significant payload in a stationary position for an operationally significant amount of time. Ensign Douglass Kondrack conducted research at AFIT where he identified that either the airship must become smaller through increased buoyancy or weight reductions or the propulsion and power systems must be made stronger for current technology to support a stratospheric airship.[98] Captain Moomey's analysis also supported the same conclusion.[99]

Payload Capacity

Payload capacity is a key capability for the WSA as it will determine the number of munitions the WSA can carry on each mission. Even if all the other capabilities are available, the WSA is not viable if it can only carry a handful of munitions to its station. A WSA that remains persistent in position at altitude over an area of interest is of no use to Joint Force Commander if it has few munitions to expend. The JFC would likely be reluctant to use the limited munitions for fear a more-important target may appear later, or may disregard the WSA altogether if a preponderance of his WSAs on station are always out of munitions.

A scan of current projected stratospheric airship technologies finds a range of payload capacities that diminish as the altitude increases. The operational version of the Lockheed Martin HAA, is projected to carry a 4000 lb. payload at 65,000 ft.[100] Future versions of the NASA Ultra Long Duration Balloon (ULDB) are projected to carry a 6000 lb. payload to 110,000 feet.[101] To carry this weight, however, the ULDB requires a volume of 631,500 m^3, well over four times the 147,000-m^3 volume of the HAA.[102] Additionally, The ULDB is also a free-floating balloon without propulsion or maneuvering systems. Other projected payload weights included the Battlelab NSMV which was projected to carry 700 lbs. to above 100,000 ft.[103] The Sanswire Stratelite is projected to carry 2000 lbs. to 65,000 feet.[104]

The above data suggests it is possible for a future WSA to carry a payload of 4000 lbs. at

65,000 feet and a payload of 2000 lbs. above 100,000 feet.

Survivability

Similar to how the balloons of World War I were a difficult target for balloon busters like Lt Frank Luke, stratospheric airships of the future will also be difficult targets. WSAs operating below 100,000 feet are within range of "double-digit" surface-to-air missiles such as the SA-10 and SA-12 as well as J-11s and other fighters; however, they would still be difficult to detect, engage, and destroy.[105,106] Due to their stationary position and lack of signatures, WSAs will be difficult to find. Even if found, they can possess simple countermeasures. Finally if successfully engaged, they will not quickly fall from their position. Unlike other weapons platforms, WSAs have the potential to be very survivable in a high-threat environment.

Stratospheric airships are inherently stealthy. Because they contain inert gas and do not produce a significant amount of heat, WSAs would present a small infrared signature at their high altitude. Because of their non-metallic structure and covering and a lack of rough edges, WSAs would also present a minimal radar return. Even with their immense size, WSAs would also be difficult to see optically at their high altitudes.[107] When detected, WSAs could also employ simple countermeasures to prevent a successful engagement. Small chaff packages could be dropped to confuse radars. Another option is to also launch numbers of inexpensive balloon decoys in the same vicinity as the WSA.[108]

Even if successful engaged by a SAM, fighter, or future directed energy weapon, WSAs are inherently survivable. WSAs would most likely use inert helium as their buoyant gas, thus there would be no flaming wreckage like would be caused by the hydrogen-filled dirigibles of the early twentieth century.[109] At operational altitudes, WSAs would have an overpressure of less than one pound per square inch. A damaged airship envelope would not cause a quick loss of pressurization and attitude, but instead slow leaks and slow descents. However, since the loss of pressure would eventually lead to a loss of aerodynamic shape required for maneuvering, a damaged WSA would need to immediately transit to a recovery location if damaged.[110] The survivability of stratospheric airships was demonstrated over Canada in 1998. Canadian F-18s were ordered to fire upon a wayward 100-meter weather balloon in order to bring it down. Even after 1000 rounds were fired at the balloon, it still managed to stay afloat for another 6 days.[111]

The biggest threat to a WSA is likely the weather. Though the winds at a WSA's operating altitude tend to be benign, seasonal or random winds could easily put a WSA far off its intended position rendering it ineffective until it could return. The lack of UV protection, solar heating, and wide temperature ranges also have the potential to create damage to an airship's fabric or payload, causing slow leaks or damaging critical components.[112]

If an SA is significantly damaged by weather or enemy action, there are options for safe return of its payload. A parachute package can be integrated into the payload. The USAF Technical Capabilities (TENCAP) program has also looked into concept of incorporating glide wings into the payload package to safely return the payload to a friendly location if a stratospheric airship is damaged.[113]

[79] Williams, "Understanding Air Density."
[80] Tomme, "The Paradigm Shift," 10.
[81] Smith, "Applications for Scientific Ballooning," 4.
[82] Ibid, 4.
[83] Ibid, 4.
[84] Lockeed Martin, "High Altitude Airship (HAA)."
[85] Tomme, "The Paradigm Shift," 49.
[86] Blackington, "Near Space Maneuvering Vehicle," 2.
[87] Ibid, 3.
[88] Tomme, "The Paradigm Shift," 47.
[89] Ibid, 47.
[90] Kondrack,"High Altitude Airship Station-Keeping," 99.
[91] Moomey, "Technical Feasibility of Loitering," 28-40.
[92] Colozza, "High Altitude, Long Endurance Airships," 5.
[93] Ibid, 1.
[94] Knoedler, "Lowering the High Ground," 117.
[95] Lockheed Martin, "High Altitude Airship."
[96] Moomey, "Technical Feasibility of Loitering," 4.
[97] Schecter, "Airships on the Rise," 30.

[98] Kondrack, "High Altitude Airship Station-Keeping," 99.

[99] Moomey, "Technical Feasibility of Loitering," 80.

[100] Schecter, "Airships on the Rise," 30.

[101] Cathay, "Development of the NASA," 13.

[102] Kondrack, "High Altitude Airship Station-Keeping," 25.

[103] Blackington, "Near Space Maneuvering Vehicle," 3.

[104] Sanswire, "Unmanned Airship Solutions," 2.

[105] GlobalSecurity.org, "S-300PMU."

[106] GlobalSecurity.org, "J-11/Su-30."

[107] Tomme, "The Paradigm Shift," 44.

[108] Roberts, "Near Space Vehicles," 19.

[109] Tomme, "The Paradigm Shift," 13.

[110] Vogt, "Performance Capability of a Damaged," 63.

[111] Tomme, "The Paradigm Shift," 13.

[112] Jamison, "High Altitude Airships," 23.

[113] Stephens, "Near-Space,"39.

Appendix B: Stratospheric Airship Small Diameter Bombs (SA-SDBs)

This appendix will describe a notional munition for use by the WSA. The WSA's munitions are the heart of the system because they provide the capability to place ordinance on target in support of CAS or TST needs. Because of its operational success, lightweight, and long range, the SDB provides an excellent baseline from which to create the WSA munition. The name of this new munition will be the Stratospheric Airship Small Diameter Bomb (SA-SDB).

The two primary needs of the SA-SDB are minimum weight and significant range. Lowering weight directly correlates to increasing the number of munitions a WSA can carry and employ. Extending range increases the radius of the area in which the WSA can engage targets. SA-SDB precision is provided by GPS-aided INS or by laser tracking. Both technologies are robust and proven in current small munition packages such as the Hellfire, Viper Strike, and SDB.

Several examples of current and proposed munitions show the SDB's weight of 285 lbs. could be dropped. Though the SMACM and other small rocket propelled munitions show potential, glide munitions provide great range without the weight required for a rocket or turbojet motor. The SSDB's weight of 80 lbs. and Viper Strike's weight of 42 lbs. show manufacture of an effective glide munition in the 80 lbs. range is possible. Another weight-saving factor would be the glide wings. The glide wings of the SDB are designed so they are prone to the weapon's exterior to minimize size for internal or external carriage on fighter aircraft. The wings then extend when the weapon launches.[114] Since the WSA payload is limited by weight, but not by size or aerodynamics of the payload, the SA-SDBs can posses glide wings already extended, saving the weight of the extension mechanism and increasing the stress-tolerances of the SA-SDB airframe.

SA-SDBs can match the range of the current SDB by exploiting the altitude of the WSA. The current range of an SDB is 85 nm. if dropped from 65,000 ft. at supersonic speed and 50nm. if dropped from 40,000 ft. at Mach 0.8.[115] WSAs will drop SA-SDBs from far higher altitudes but with no initial airspeed. This paper assumes the SA-SDB can achieve a gradual pull-up from vertical drop to horizontal glide without sacrificing significant airspeed. Using this assumption, which costs an extra 15,000 feet of altitude, an SA-SDB can achieve a velocity exceeding Mach 0.8 following a 25,000-ft. drop from WSA altitudes and can exceed Mach 1.5 following a 45,000-ft. drop. In effect, the SA-SDB will achieve a glide at 0.8 Mach by the time it is at 45,000 feet to achieve a 50-nm. range. Similarly, an SA-SDB dropped from over 100,000 feet can achieve Mach 1.5 before it reaches 65,000 feet thus enabling the SDB-like higher-speed, higher altitude range of 85nm. The anticipated profiles of these munitions drops are displayed in Figure B-1.

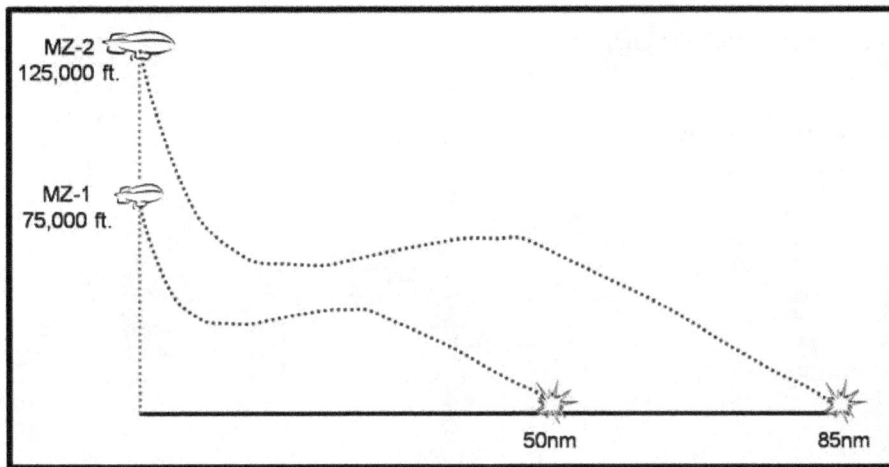

Figure B-1: Drop Profiles of SA-SDBs from 75,000 ft. and 110,000 ft.

Several considerations will need to be taken into account when designing a future SA-SDB. The first is temperature. The current SDB is rated for -40°F.[116] The SA-SDB will need to be designed with better materials or heaters to withstand high altitude temperatures that can range as low as from -75°F at 65,000 ft., increasing to -10F at 130,000 ft.[117] The current SDB is a 3G airframe.[118] The stresses of a drop from high-altitude, increasing velocity, and need to gradually pull out of the drop will likely require an increase in SA-SDB airframe strength. The current SDB is also limited by battery life once dropped.[119] The SA-SDB design will need to account for the additional time of flight required for drops from a WSA.

The primary issue of an SA-SDB supporting CAS or TST missions is time-of-flight. The current SDB, when dropped from 40,000 ft. at Mach 0.8, requires 10 minutes, 20 seconds to reach a target at 40 nm. range.[120] Simple extrapolation suggests a 13-minute time-of-flight to targets at 50nm and a 22-minute time-of-flight to targets at 85 nm. An additional one to two minutes for the velocity drop from high altitude must also be factored into these flight times. Though the WSA and SA-SDB provide persistence, significant range and flexibility for CAS and TST missions, the time-of-flight of the munitions will create issues for moving or fleeting targets.

---

[114] Boeing, "Small Diameter Bomb Increment I."
[115] Vogel, Email Dated 7 Nov 2008.
[116] Ibid.
[117] Tomme, "The Paradigm Shift," 39.
[118] Vogel, Email Dated 7 Nov 2008.
[119] Ibid.
[120] Ibid.

Appendix C: CAS/TST

This appendix will provide additional information on the Close Air Support (CAS) and Time Sensitive Target (TST) missions.

CAS

The key member of a CAS mission is the Joint Terminal Attack Controller (JTAC). Joint Doctrine defines the JTAC as "a qualified (certified) service member who, from a forward position, directs the action of combat aircraft engaged in close air support and other offensive air operations."[121] JTACs typically reside in the Air Support Operations Center (ASOC) collocated with the ground element's tactical command post. They can also accompany front line ground forces within visual sight of CAS targets. JTACs identify targets and pass targeting information to CAS aircraft supporting the ground element and thus would be the personnel to direct CAS fire from a supporting WSA.

JTACs can have three different types of control: Type 1, where both the JTAC and CAS aircraft have visual acquisition of the aircraft; Type 2, where the JTAC does not have visual acquisition of the aircraft and/or the target; and Type 3, where the JTAC provides clearance for multiple attacks on a target, set of targets, or geographic area.[122] Since the airship would be unmanned and out of visual range of the JTAC, Type 2 control would exist in WSA CAS engagements. The JTAC could control the engagement via satellite communications (SATCOM) with the airship operator, or if properly trained and certified, directly control munitions drop through a direct HF data link to the airship.

TST

The second type of targeting situation where a weaponized stratospheric airship would prove useful is time-sensitive-targets (TSTs). JP 3-60, Joint Targeting, asserts TSTs require immediate response because they are "highly lucrative, fleeting targets of opportunity." or they present immediate danger to a JFC's forces.[123] Efforts over the past two decades to effectively engage TSTs are rooted in experiences during Operation Desert Storm where the US had difficulty finding and engaging Scud missile launchers. Since Scud attacks on Israel had the potential for major coalition implications, the US expended an inordinate amount of time and resources to find and destroy these fleeting targets. Enemies in more recent conflicts have attempted to limit target vulnerability including the Serbs hiding tanks in Kosovo and Al Qaeda and Taliban forces hiding in Afghanistan caves.[124]

The US has endeavored to effectively engage TSTs by compressing the Find, Fix, Track, Target, Engage, and Assess (F2E2EA) cycle to less than 10 minutes. Several successful efforts have compressed the command and control (C2) actions associated with TST targeting; however, improvements for TST engagement are still needed. A quote from Lieutenant General Daniel Leaf, when he was the Director of USAF Operational Requirements, highlights a shortfall in the TST goal: "If an airplane is 20 minutes away from a target, all the data links in the world are not going to make the kill chain nine minutes." [125] This lack of persistence led to USAF efforts adding precision munitions to UAVs. Commanders can now engage TSTs from the same ISR platform that detected the TST. This capability has led to several TST successes in Iraq Afghanistan.[126] This success could be carried forward with WSAs.

---

[121] JP 3-09.3, "JTTPs for CAS." GL-12.
[122] Ibid, V-14.
[123] JP-3.60, "Joint Targeting," I-5.
[124] Hebert, "Compressing the Kill Chain," 50.
[125] Ibid, 53.
[126] Ibid, 54.

Abbreviations

C2	Command and Control
CAS	Close Air Support
CONOPS	Concept of Operations
CONUS	Continental United States
DARPA	Defense Advanced Research Projects Agency
DE	Directed Energy
DOD	Department of Defense
EO/IR	Electro Optical / Infrared
HAA	High Altitude Airship
IADS	Integrated Air Defense System
ISR	Intelligence, Surveillance and Reconnaissance
MDA	Missile Defense Agency
MZAHC	MZ Airship Handheld Controller
NASA	National Aeronautics and Space Administration
SA	Stratospheric Airship
SA-SDB	Stratospheric Airship Small Diameter Bomb
SAM	Surface-to-Air Missile
SDB	Small Diameter Bomb
SIGINT	Signals Intelligence
SMDC	Army Space and Missile Defense Command
SSDB	Short Small Diameter Bomb
TIC	Troops in Contact
TST	Time Sensitive Target
UAV	Unmanned Aerial Vehicle
ULDB	Ultra Long Duration Balloon
WSA	Weaponized Stratospheric Airship

Bibliography

Absolute Astronomy.com. "Cumulonimbus Cloud," http://www.absoluteastronomy.com/topics/Cumulonimbus_cloud.

Air Force Instruction (AFI) 16-401. *Designating and Naming Defense Military Aerospace Vehicles,* 14 April 2005.

Allen, Edward H. "The Case for Near Space." *Aerospace America* 44, no. 2 (2006): 31-34.

Amol, Sharma. "Floating a New Idea for Going Wireless: Parachute Included." *Wall Street Journal.* 20 February 2008.

Belote, Howard D. "Counterinsurgency Airpower: Air-Ground Integration for the Long War." *Air and Space Power Journal* 20, no. 3 (Fall 2006): 55-68.

Blackington, Robert. "Near Space Maneuvering Vehicle." AIAA Presentation. Responsive Space Conference. 2003.

Bolkom, Christopher. *Potential Military Use of Airships and Aerostats.* (Library of Congress: Congressional Research Service, 2005) 4.

Borden, Steve. "Viper Strike." Army Program Executive Officer, Missiles and Space, April 2006.

Brown, David R. "JTAC: MOA vs. JTTP." *Field Artillery Journal*, January-February 2005: 18-21.

Cathey, Henry M. Jr.and David L. Pierce, "Development of the NASA Ultra-Long Duration Balloon." NASA Science Technology Conference 2007.

Chu, Adam, Mo Blackmore, Ronald G. Oholendt, Joseph V. Welch, Gil Baird, David P. Cadogan, and Stephen E. Scarborough. "A Novel Concept for Stratospheric Communications and Surveillance: Star-Light." White Paper. Colorado Springs, CO: Near Space Systems Inc., 2006.

Colozza, Anthony and James L.Dolce. "High-Altitude, Long-Endurance Airships for Coastal Surveillance." White Paper. Brook Park, OH: NASA STI Program Office, 2005.

Colozza, Anthony. "Initial Feasibility Assessment of a High Altitude Long Endurance Airship." White Paper. Brook Park, OH: NASA STI Program Office, 2003. http://gltrs.grc.nasa.gov/reports/2003/CR-2003-212724.pdf

Fisher, Jr., Richard. "Deterring a Chinese attack against Taiwan: 16 steps." White Paper. Alexandria VA: International Strategy and Assessment Center, 2 April 2004.

GlobalSecurity.org. "Chengdu J-10 (Jian-10 Fighter aircraft 10) / F-10." http://www.globalsecurity.org/military/world/china/j-10.htm.

GlobalSecurity.org. "J-11 [Su-27 FLANKER] Su-27UBK / Su-30MKK/ Su-30MK2." http://www.globalsecurity.org/military/world/china/j-11.htm.

GlobalSecurity.org. "S-300PMU / SA-N-6 SA-10 GRUMBLE." http://www.globalsecurity.org/military/world/russia/s-300pmu.htm#prc.

GlobalSecurity.org. "S-300V / SA-12A GLADIATOR and SA-12B GIANT HQ-18." http://www.globalsecurity.org/military/world/russia/s-300v.htm.

Gordon, Walter O., and Chuck Holland. "Back to the Future: Airships and the Revolution in Strategic Airlift." Air Force Journal of Logistics 29: 3/4 (Fall/Winter 2005): 46-56.

Haffa, Robert T., and Jasper Welch. "Command and Control Arrangements for the Attack of Time Sensitive Targets" White Paper. Northrop Grumman Analysis Center. 2005.

Haun, Phil M. "The Nature of Close Air Support in Low Intensity Conflict." *Air and Space Power Journal* 20: no. 3 (Fall 2006): 107-110.

Hebert, Adam J. "Compressing the Kill Chain." *Air Force Magazine* 86:03 (2003): 50-54.

"High Altitude Airship (HAA): Global Persistent ISR" Brochure. Akron, OH: Lockheed Martin Missiles and Fire Control, 2008.

Ingram, Bernd L. "Joint Targeting for Time-Sensitive Targets." *Field Artillery Journal,* May – June 2001: 28-31.

Jamison, Lewis, Geoffrey S. Sommer, and Issac R. Porche III. "High-Altitude Airships for the Future Force Army" Rand Corporation Technical Report. 2005.

Joint Publication (JP) 3-09.3. *Joint Tactics, Techniques, and Procedures for Close Air Support (CAS)*, Incorporating Change I, 2 September 2005.

Joint Publication (JP) 3-09.34. *Multi-Service Tactics, Techniques, and Procedures (MTTP) for Kill Box Employment*, 14 June 2005

Joint Publication (JP) 3-60. *Joint Targeting*, 13 April 2007.

Knight, Anthony G. and Aaron B. Luck. "Tactical Space – Beyond Line of Sight Alternatives for the Army and Marine Corps Ground Tactical Warfighter." Research Thesis. Monterey, CA: Naval Postgraduate School, 2007.

Knoedler, Andrew J. "Lowering the High Ground: Using Near-Space Vehicles for Persistent ISR." Research Paper. Maxwell AFB, AL: Air Command and Staff College, 2005.

Kondrack, Douglas P. "High Altitude Airship Station-Keeping Analysis." Research Thesis. Wright-Patterson AFB, OH: Air Force Institute of Technology, 2005

Lake, James P. "Continuously Available Battlefield Surveillance." Research Paper. Maxwell AFB, AL: Air Command and Staff College, 2007.

"Laser Focused Lethality Munition (LFLM)." White Paper. St. Louis, MO: Boeing Integrated Defense Systems, 2008.

Marilao, Frank Q. "Dirigibles of Death - Awakening a Failed RMA: An Analysis of the Potential Threat Posed by Unmanned Airships of the Future." Research Paper. Maxwell AFB, AL: Air Command and Staff College, 2008.

Mayer, Norman. "Lighter-Than-Air Systems," *Aerospace America*, December 2005, 30.

Miller, Tim and Mathias Mandel, "Airship Envelopes: Requirements, Materials and Test Method" White Paper: Frederica, DE: ILC Dover, Inc.

Milliron, Steven P. "Army JTAC Training – The Way Ahead." *Field Artillery Journal,* March-June 2004: 50-54.

Minnick, Wendell. "Chinese Air Power Focuses on Taiwan, U.S. Scenarios." *Defense News.* 14 July 2008.

Moomey, Eric R. "Technical Feasibility of Loitering Lighter-than-air Near-space Maneuvering Vehicles." Research Thesis. Wright-Patterson AFB, OH: Air Force Institute of Technology, 2005

"Next-Generation Aviation Missile: SMACM: Surveilling Miniature Attack Cruise Missile" Brochure. Orlando, FL: Lockheed Martin Corporation. 2006.

"Northrop Grumman to Build Stratospheric Surveillance Airship." *Military & Aerospace Electronics Magazine,* 11 May 2006.

Pirnie,Bruce R., Alan Vick, Adam Grissom, Karl P. Mueller, and David T. Orletsky. "Beyond Close Air Support: Forging a New Air-Ground Partnership" Rand Corporation Technical Report. 2005.

Roberts, Garren B. "Near Space Vehicles: Improvements for the Near Future." Research Paper. Maxwell AFB, AL: Air Command and Staff College, 2006.

Schanz, Marc V. "The Airpower Surge." *Air Force Magazine* 92, no. 1 (2009): 28-32.

Schecter, Eric. "Airships on the Rise - Blimps to Challenge UAVs as ISR Craft." *C4ISR – The Journal of Net-Centric Warfare*, September 2008: 30.

Schmidt David K., James Stevens, and Jason Roney. "Near-Space Station-Keeping Performance of a Large High-Altitude Notional Airship." *Journal of Aircraft* 44, No. 2 (March–April 2007): 611.

"SDB Focused Lethality Munition." Background Paper. St. Louis, MO: Boeing Integrated Defense Systems, 2008.

"Small Diameter Bomb Increment I." Background Paper. St. Louis, MO: Boeing Integrated Defense Systems, 2008.

"Small Diameter Bomb Increment II." Background Paper. St. Louis, MO: Boeing Integrated Defense Systems, 2008.

Smith, Michael S., and Edward E. Rainwater. "Applications of Scientific Ballooning Technology to High Altitude Airships," Raven Industries Inc., Sulphur Springs, Tex., Presented to the American Institute of Aeronautics and Astronautics conference. Denver, Colo., 2003.

Stephens, Hampton. "Near Space." *Air Force Magazine* 88, no. 7 (2005): 36-40.

Steves, Mark. "Near-Space 2015: A Conceptual Vision of Near-Space Operations" *Air and Space Power Journal* 20, No. 2 (Summer 2006): 110-117.

Thill, Joseph T., "Penetrating the Ion Curtain: Implications of Directed Energy Integrated Air Defense Systems in 2030." Research Paper. Maxwell AFB, AL: Air Command and Staff College, 2008.

Tomme, Edward B. "The Paradigm Shift to Effects-Based Space: Near-Space as a Combat Space Effects Enabler." Airpower Research Institute Research Paper 2005-01. Maxwell AFB, AL: CADRE, 2005.

Tomme, Edward B., and Sigfred Dahl. "Balloons in Today's Military – An Introduction to the Near-Space Concept." *Air and Space Power Journal* 19, no. 4 (Winter 2005): 39-49.

Torbinson, Eric. "Lockheed Martin Makes a Defensive Strike for the F-35." *Dallas Morning News*. January 6, 2009.

"Unmanned Airship Solutions for Integrated ISR and Communications Systems" Technical White Paper. Sanswire Corp. 2007

US Army. "HA: High Altitude Efforts." Huntsville, AL: Army Space and Missile Defense Command/Army Forces Strategic Command.

US Air Force. "Fact Sheet: GBU-39B Small Diameter Bomb Weapon System." Langley AFB, VA: Air Combat Command, November 2007.

US Air Force. "Fact Sheet: F-16 Fighting Falcon." Langley AFB, VA: Air Combat Command, October 2007.

US Air Force. "Fact Sheet: F-22 Raptor." Langley AFB, VA: Air Combat Command, April 2008.

US Air Force. "Fact Sheet: RQ-4 Global Hawk Unmanned Aircraft System." Langley AFB, VA: Air Combat Command, October 2008.

US Air Force. "Fact Sheet: Tethered Aerostat Radar System." Langley AFB, VA: Air Combat Command, August 2007.

US Department of the Interior. "Mineral Commodities Summaries" Washington DC: US Geological Survey, January 2008.

US Department of State. "Background Notes: Taiwan." Washington DC: Under Secretary for Public Diplomacy and Public Affairs, September 2008.

Vogel, Michael R. E-mail dated 7 November 2008 from Boeing Integrated Defense Systems.

Vogt, Jr., Charles W. "Performance Capability of a Damaged Lighter-than-Air Vehicle Operating in the Near Space Regime." Research Thesis. Wright-Patterson AFB, OH: Air Force Institute of Technology, 2006.

Walker, Mary. "Memorandum for the Secretary and the Chief of Staff of the Air Force, Subject: Legal Regime Applicable to 'Near Space'." 27 September 2004.

Williams, Jack. "Understanding Air Density and Its Effects." *USAToday*, 17 May 2005. http://www.usatoday.com/weather/wdensity.htm

Wilson, Jim. "21st Century Airships are the Ultimate Weapon in the War on International Terrorism." *Popular Mechanics*. March 2002.